Ham Radio Exam Prep

A License Manual and Study Guide for the Amateur Radio General Class and Radio Technician Tests with 100 Test Questions

Copyright 2020 by Ham Radio Team - All rights reserved.

This book is geared towards providing precise and reliable information about the topic. This publication is sold with the idea that the publisher is not required to render any accounting, officially or otherwise, or any other qualified services. If further advice is necessary, contacting a legal and/or financial professional is recommended.

-From a Declaration of Principles that was accepted and approved equally by a Committee of the American Bar Association and a Committee of Publishers' Associations.

In no way is it legal to reproduce, duplicate, or transmit any part of this document, either by electronic means, or in printed format. Recording this publication is strictly prohibited, and any storage of this document is not allowed unless with written permission from the publisher. All rights reserved.

The information provided herein is stated to be truthful and consistent, in that any liability, in terms of inattention or otherwise, by any usage or abuse of any policies, processes, or directions contained within, is the solitary and utter responsibility of the recipient reader. Under no circumstances will any legal responsibility or blame be held against the publisher for any reparation, damages, or monetary loss due to the information herein, either directly or indirectly.

Respective authors own all copyrights not held by the publisher.

The information herein is offered for informational purposes solely, and is universally presented as such. The information herein is also presented without contract or any type of guarantee assurance.

The trademarks presented are done so without any consent, and this publication of the trademarks is without permission or backing by the trademark owners. All trademarks and brands within this book are thus for clarifying purposes only and are the owned by the owners themselves, and not affiliated otherwise with this document.

Published by Newstone Publishing

ISBN 978-1-989726-04-4 (paperback)

Table of Contents

Introduction 14

Chapter 1 – Knowing Ham Radio 15

    From Then to Today 16

    Why Join Ham Radio – Things Hams Do 17

        Operations 17

        Technology 18

        Joining the Community 19

    Amateur Radio Activities 20

        Making Contacts and the Identification Process 20

        Novel Activities 25

Chapter 2 - Getting Acquainted with the Technology 26

    Modulation and its Types 28

    Basics of Electricity 33

    Advanced Electronic Components 37

    Understanding Basic Radio Gadgetry 40

Chapter 3 – Handle Your Equipment Well 44

    A. Basic Test Kit 44

    B. Requisites for Building and Repairing 47

    C. Tools for Maintenance 50

    D. Spare Parts, Adapters and Common Consumables 52

    E. How to Maintain your Ham Radio Station 54

    F. Competent Troubleshooting 57

    G. Troubleshoot Your Ham Radio Station 58

    H. Dealing with Interference 66

    I. Rules for Part-15 Appliances 74

  J. Build/Not to Build 74

Chapter 4—How Do Radio Signals Propagate 76

  How Do Radio Signals Travel: Propagation 76

  Antennas Play a Key Role in Propagation 79

Chapter 5—Transmission and Reception of Radio Signals 83

  The Band, the Frequency and the Mode 83

  Functions of the Transmitter 84

  Functions of a Receiver 85

  Digital Form of Communication in Amateur Radio 87

  Use of Batteries and Power Supplies in Amateur Radio Services 91

  Radio Frequency Interference (RFI) 94

  The RF Grounding Rules for the Amateurs 99

Chapter 6—Finding Your Friends; Finding Other Hams 102

  Radio Clubs 102

  The Relay League of America 103

  Ham Clubs, Specialty Organizations and Online Communities 105

  Ham Festivals and Conventions 107

  DXing 108

  Making a Contact 112

  Ragchewing Etiquettes 124

Chapter 7—Build a Home for your Ham Radio 129

  Getting the Ergonomics Right 129

    The Operator's Chair 130

    The Operating Desk and Shelves 131

  Basic Equipment 132

    Purchasing New/Used Equipment 134

Placement of Gear	135
The Computer	136
Placement of Computer	138
Lighting	139
Feed lines	140
Maintaining a Notebook	141
The Log Book	141
The Radio	142
Taking Charge of Radio Frequencies and Electrical Safety	143
Rules and Regulations	143
Handling DC	144
Family Safety	144
Lightning Strikes	144
The Human Body and RF	146
First-Aid Kit	146
Grounding Power	147
Grounding RF Signals	148
Basic Ground System	149
Safety Grounds	151
Ground for RF Return Path	152
Induced Ground Currents	153
RF Exposure	154
Décor	159
Upgrade Your Shack	161
Chapter 8—Let Your Radio Announce Your Airy Presence	163
A. Goals for Setting up the Radio Station and Usage of Personal Resources	163

- Working from Home 163
- Going Mobile 164
- Going Portable 164
- Tackling Handheld Radios 165

B. Tips for Selecting a Radio 166
- Shortwave/HF Radios 167
- Basic HF Radio 167
- Journeyman 167
- High Performance 168
- Select a Filter 168
- HF Radios for Mobile and Portable Operators 169
- Digital Data for High Frequencies 170
- Deciding on Amplification 171
- All about UHF and VHF Radios 172
- Radios with FM Only 173
- Mobile Radios 174
- UHF/VHF Amplifiers 175

C. Tips for Selecting an Antenna 176
- Antennas for High Frequency 176
- Portable and Mobile Antennas 184
- Feed Lines and Connectors 186

D. Support for the Antenna 191
- Masts and Tripods 191
- Towers 192
- Trees 195
- All About the Rotator 196

7

  E. Accessories for Your Radio 198

   Microphones 198

   Keys and Keyers 199

   Antenna Tuners 200

   Controlling the Radio 201

   The Digital Mode 201

Chapter 9—The Licensing Regulations and System 203

  An Overview of the Amateur System 203

  Types of Licenses 205

  Licensing Examinations 208

  Preparing for the Examination 210

  License Renewal Process 212

  Responsibilities as Part of Licensed Ham Radio 212

  Privileges and Bands 213

  International Rules and Operating 216

Chapter 10—Operating Regulations 218

  Guest Operating 218

  Identification Rules and Process 219

  Interference in Transmissions 223

  Different Types of Transmissions and Communications 225

   Third-Party Communications—The Hows and Whys 225

   Prohibited Transmissions 227

  QSL Cards 230

Chapter 11—Use Communication Skills to Share and Care 231

  A. Linking Up with an Emergency Services Establishment 231

   ARRL 232

  ARES and RACES... 235

  MARS 236

 B. Preparing for Participation in Emergency Operations 237

  WHO... 237

  WHERE 238

  WHAT 238

  Go-Kit 238

  HOW... 240

 C. Conducting Operations During Emergencies 240

  Make a Report 241

  Use an Auto Patch 243

  Make a Distress Call 245

  Render Support to Outside Communications 247

 D. Offering to Serve the Public... 247

  Monitor the Weather 250

  Attend Sporting Events and Parades 250

 E. Net Operations 251

 F. The Handling of Traffic 254

  Movement of Traffic.. 255

  The Radiogram... 255

  Deliver the Message 258

Chapter 12—Some Handy Tips for Amateurs 260

 A. When you Begin your Journey 260

  You are Not a Know-it-All 260

  Be a Good Listener... 260

  Be Courteous 260

- Find a Buddy 261
- Join In 261
- Befriend your Equipment 261
- Listen to the Manufacturer 261
- Risk Experimenting 262
- Never Let Hiccups Overcome You 262
- Be a Relaxed Operator 262

B. Listen to the Experts 263
- Learning is a Continual Journey 263
- Be a Great Listener 263
- Details are Important 263
- Antennas are Important 264
- Comprehend the Decibels 264
- Know Your Equipment 264
- Practice, and Practice Some More 265
- Regularly Sharpen Your Brain 265
- Appreciate Ham Radio History 265
- Do Things the Right Way 266

C. Operating Your First Ham Radio Station 266
- Getting Used to Used Equipment 266
- Construct Your Own 267
- Build Your Own Cables 267
- Progress Step-by-Step 267
- Look and Learn 267
- Distribute Your Budget 268
- Keep Track of the Weakest Link 268

    Be Flexible 268

    Be Comfortable 269

    Be Well Grounded 269

D. Having Fun on the Radio 269

    Listen, and Find Fun 270

    Unique Contests and Events 270

    Create Your Own Contest 270

    Join the Parade 271

    Make New Friends 271

    Send a Radiogram 271

    Contact Through Satellite 272

    SWL-ing 272

    Pick Up a New Language 272

    Travel to a Cool Place 273

E. Returning the Favor to the Ham Radio Community 273

    Be a Tester of Diverse Products or a QSL Manager 273

    Engage in On-the-Air Monitoring 274

    Be a Representative of Ham Radio 274

    Offer to Experiment 274

    Be an Elmer 275

    Be a Volunteer 275

    Offer Public Service Assistance 275

    Preparedness for Emergencies 275

    Community Prepares for Emergencies 276

    Nurture Long-Term Friendships 276

The Question Pool 277

The Answers 301

Introduction

Ham radio, or amateur, is not something that has been recently invented or discovered; it has been around for quite some time. But it has undoubtedly emerged, transformed, branched and amplified itself over the years. Known as the most powerful means of communication available to human beings, it is practiced by people in different age groups, ranging from preschoolers to senior citizens. The siren call of this radio can attract those who have deep expertise in the technical domain as well as those who have never held a microphone in their life. You might have seen these amateur radios being used in movies or books, or maybe by a hobbyist friend.

While some hams are focused on the science or technology used in radio, some are used to keep in touch with others. While conventional shortwave bands are still surrounded by ham signals, traversing all around the planet, today's modern hams transmit data through airwaves, using different kinds of transmitters such as internet, microwave transmitters and lasers, which can travel to unusual places.

To keep businesses or organizations from exploiting the amateur band, amateurs are prohibited from receiving compensation for what they do. For instance, you cannot accept a fee if you provide communication for a charity. This ensures the amateur band is not exploited and is free to be improved upon.

In simple words, ham radio offers the most powerful means of wireless communication available. Let's look deeper and see what ham radio can do for us.

Chapter 1 – Knowing Ham Radio

You might have an image in mind for a ham radio based on what you have seen in a movie or newspaper. But there are different types of hams, ranging from casual chatters to ones used in emergency conditions. They use different types of antennas on a variety of frequencies to communicate with hams across the city and around the globe. People use ham radio for keeping in touch with their family, for personal enjoyment, for experimenting with radio equipment and for emergency communications. For communication purposes, they use devices such as computers, lasers, telegraph, microphones, cameras and even satellites.

As mentioned earlier, commercial ham activity is banned. Some traditional bands are really active worldwide. On the other hand, other bands are best suited for performing public regional/local communication.

As they are granted access to various frequency bands across the spectrum, hams use a variety of frequencies, allocated by the FCC. They might operate from a band that's just above the AM broadcast, all the way to the microwave region, which is the gigahertz range. There are many bands that are found in the range that goes just above the AM radio band to above the citizen band. The bands being used also depend on what time of the day the communication is scheduled to occur. For instance—during the day, 15 to 27 MHz can be used for long-distance communication, whereas at night 1.6 MHz to 15MHz works out well.

There are different ham radio organizations and clubs, created for different purposes. There are even host conventions and ham flea markets devoted to hams—some of whom come from a technical background. But one common thing about all these diverse hams is that they are interested in radio.

A teen sitting in Miami can make friends over the airwaves with a ham in Singapore. An aircraft engineer sitting in London can participate in an annual contest with other hams in 100 countries. In Indonesia, volunteers can pass on messages in the aftermath of a volcano. Thus, it is clear that friendship, emergency services and convenience are some of the main attributes of any ham radio.

From Then to Today

Soon after Marconi roamed the Atlantic in 1901, some curious people started experimenting with wireless amateur radio, which led to the invention of broadcasting and wireless telegraphy. The first amateur radio license was granted in 1912, and since then the number of hams has grown exponentially. Back in the day, stations used a noisy and vigorous arc, called a spark, to generate radio waves. However, these dangerous and inefficient sparks were soon replaced by efficient vacuum-tube transmitters, and by the end of 1920, voice code could be heard on the airways. Instantly connecting different parts of the world, individuals and communities across the globe, radio became really popular.

With the growing popularity of radio communication, there emerged the need for a central body to regulate its competing uses, such as military communications, broadcasting, news services and public safety. Thus, the Federal Communications Commission (FCC) was formed.

The amateur radio was introduced in 1934. Experts skilled in radio communication played a crucial role in World War II as radio engineers. To meet the growing needs of the world, more and more hams chose communication and radio as their profession during the 1950s and 1960s. This led to the evolution of the communication industry all around the world. The amateur airwaves were filled with voice signals, and even a new form of picture transmission was invented that utilized regular voice receivers and transmitters. In the 1970s, regional communications built repeater stations which provided low-power mobile and handheld radios. Then, in the 1980s and 1990s, microprocessors were applied to radios, increasing the capabilities of ham equipment in the world of digital communications.

Hams also developed packet radio, which was an adaption of computer network technology and is now used for public safety and commercial communications. More capabilities were added to these amateurs by personal computer, which enabled recordkeeping, designing and modeling. With the advent of the internet, hams accordingly adapted the new technology to suit their respective uses.

Today, in the digital era, where wireless is still at the forefront of communication technology, ham radio continues its journey of innovation by combining the power of internet and communication. Hams use various types of signals such as voice signals and Morse code. Others design their own protocols to send messages and data around the world. Today we have separate wireless data networks exclusively for hams and even a radio-based email network that enables the ham to log in from any corner of the globe. To connect hams in different parts of the globe using handheld transmitters, voice communications utilize radio links and the internet. Some hams assemble their own communication stations to transmit high-quality videos. Amateurs also communicate with each other using complicated radios that help them connect directly.

In addition to all this, hams also look at the skies for communication purposes—there are several amateur satellites that link hams on the ground with the help of data signals, Morse code and voice. Unlike frequencies used by AM and FM radio stations, which are limited to 40-50 miles, as they are line-of-sight, the software used by hams lets these signals bounce off the moon and meteor trails present in the space. Natural disasters disrupt normal communication systems and that's when hams pitch in to help. Therefore, ham radios form an important part of many disaster relief plans. They also help in providing communication for festivals, sporting events, parades and so on.

Why Join Ham Radio – Things Hams Do

There are three main aspects of ham radio that hams are interested in—operations, technology and social. But no matter what a ham's particular interest, they are required to have a radio license.

Operations

When you tune a radio to a ham frequency, you'll hear hams talking about different things. Let's take a look at some of the common types of activities they are generally involved in.

Ragchews – The most common types of activity hams are involved in is conversational. This kind of light chatter is known as 'chewing the rag,' and therefore the hams involved in this process are known as 'ragchews.' Having a conversation over ham radio is quite simple. Just make contact and get started.

Nets – Nets, or networks, are on-the-air meetings set up for hams who share similar hobbies or areas of interest. Some common nets include technical service nets, in which stations call with specific problems or questions; traffic nets that move traffic or text messages via ham radio; swap nets that help hams to sell items or list things they need, etc.

Contests, awards and DXing (DX stands for distance) – Hams compete with each other to contact faraway stations, seeking out people living on islands and in foreign countries. When conditions allow, the ham radio band is full of people from different parts of the world. Participating in contests, hams try to make as many contacts as possible, sending and receiving different messages related to the contest that is being held, related to a certain area, using a specific band. In addition to these contests, hams make contact whenever there are special events or award ceremonies for various accomplishments. For example, a station was set up in North Carolina when the Wright Brothers took their first flight.

Technology

Receiving and transmitting radio signals is an electronics-intensive effort. Ham radio is comprised of everything from analog electronics to the latest digital signal processing and computing. Experts also choose to design and build their own stations and tools using factory-built components that are widely available in stores, on and offline. This is known as homebrewing, and hams often help each other build their own stations.

Sometimes hams also create their own software and use the internet in tandem with radio to create efficient hybrid systems. By adapting different data transmission protocols over computer networks, they create packet radio, which is useful for various commercial purposes. By combining GPS technology and the power of internet, the Automatic Position Reporting System was developed, which is widely used in various fields today. Morse code communication is still a well-known technology that hams use to interact with each other, but digital operations are emerging at a faster speed.

One of the most common home stations today is a combination of radio and computer, using software-defined technology that helps reconfigure radio signals under software control. Apart from computers, hams are also interested in propagation and antennas, which involves radio signals traversing from place to place. They are interested in sunspots, solar cycles and how these impact the ionosphere of our planet. In all this, weather plays an important role as to how long a radio signal can travel.

Experimenting with antennas is something hams are really interested in. They come up with new designs and contribute to different refinements created by those who design the art. These systems of antennas can be small or as big as towers supported with rotating arrays.

Radio technology is used in meteorology, rocketry and radio control. Hams have special frequencies for these purposes that are separate from the crowded unlicensed radio control band. Ham links are also used in aviation, rallies, auto racing and so on. Whatever the aspect of computing and electronic technology, it can be used in ham radio.

Joining the Community

Due to ham radio's simple infrastructure, amateurs can easily tune in to help in cases of emergency or disaster when communication over normal channels becomes difficult. They organize for help in a variety of emergency conditions to support safety agencies. And whenever there is any forecast of a disaster, such as a hurricane about to hit a country, hams gear up to communicate from the affected areas so that quick help can be offered to the regions. They coordinate with rescue and safety agencies and handle health and welfare messages to support all those working in this state of emergency.

There are several disabled hams who offer their unique talent, although they cannot be physically present. There are organizations that help those with disabilities obtain a ham license, providing the necessary instructions and material. A ham license is a license to learn how to transmit your signals and make contacts over the air with other hams. Just like a pilot or driver's license, you improve with practice. And as you gain more experience, you can upgrade your license to get permission to try out more interesting things.

Ham radio broadcasts in all directions, just like normal radio, but hams generally do not use their radio in a broadcast manner like a radio jockey does. In the case of a normal FM or AM radio, one person speaks and the signals are transmitted to thousands of people who are tuned in to listen. On the other hand, hams conduct a two-way communication with another ham, or a group of hams, in an informal manner. These hams can be in the same town, state, county, country or continent, or can be in different countries, based on what frequency is being used for communication. As previously discussed, they also participate in nets, exchanging messages at predetermined frequencies and times.

Many hams get started with VHF using battery-operated handheld transceivers to transmit on one frequency, while receiving on another. They use FM repeaters, which are organized and supported by radio clubs and organizations. There are several such FM repeaters across the country. These repeaters use common receive-and-transmit frequency pairs to borrow antenna space from TV station owners. When the repeater receives a signal, it rebroadcasts it to another frequency using more power than what is available from handheld radio. This way the range of signal coming from handheld radio is extended from a few miles to several hundred.

Amateur Radio Activities

As we saw in the previous section, hams are involved in lots of things. Certain things they do are quite confusing. That's why it is good to have an understanding of the basics of ham radio activities or the fundamentals that are part of all ham radio communications.

Making Contacts and the Identification Process

Each ham is identified on the airwaves by a call sign. Instead of using normal names such as John or Stake, you might hear "John K1MMH" or "Stake G4BUP." This call sign makes you completely unique among all ham users across the globe. There will only be one K1MMH or one PP5GR. By stating this call sign, you let other hams know who you are and where you are located. Identifying yourself using the call sign is known as signing, and this is important since you cannot see the other ham, except in the case of video transmissions.

Any conversation between hams over the air is termed a contact, and when the conversation starts, it is known as making a contact. You are required to send your call sign regularly, at the beginning of contact, during the contact and before closing the contact, so that everyone consistently knows who they are communicating with. When you transmit your call sign in order to make a contact, it is called calling. When you are making a "come in anybody" call, where anyone can respond, this is called a general call or a CQ.

Once the contact is successfully made, the next step is to share some more important information, such as a signal report—something which helps other stations understand how well you are receiving their information. Then you exchange name, location and other important attributes. Once all this is done, you proceed with transmitting the actual information and at the end, you sign off by stating your call sign. However, there are several ways to exchange information:

Ham shorthand: Just like flying, radio has its own terminology which it has been using for years. In fact, some of these conventions date back to the days of telegraph, when every individual word needed time to be transmitted/received. Therefore, the operators came up with a series of characters and abbreviations that ensured quick and smooth transmission. For instance, the word "break," which was originally used to refer to disconnecting the telegraph line, so that no characters were transmitted. Using a similar strategy, modern-day operators began using shorthand language, known as "Q-signals." for example—instead of saying "what is your location?" ham operators use "QTH."

The system is comprised of a set of standard abbreviations used to communicate basic words and information, saving time and allowing easy communication between operators who are not able to communicate due to the lack of a common language. Amateurs today use Q-signals extensively for communication purposes. The following table lists some of the commonly used Q-signals. While these were actually developed to be used by Morse operators, they are also used on the phone. When you hear something like, "QRZed?", it means "who is trying to call me?"

Abbreviation	Description
QRG	Will you tell me my (or your) exact frequency? Your exact frequency is ... kHz.
QRL	Are you busy? I'm busy with ... Generally used to see if a given frequency is busy.
QRM	Is my transmission being interfered with? Your transmission is being interfered with ... 1. Nil, 2. Slightly, 3. Moderately, 4. Severely, 5. Extremely

QRN	Are you hearing static?
	I've got static … (fill in with the 1-5 value under QRM).
QRO	Shall I increase the power?
	The power should be increased.
QRP	Shall I decrease the power?
	The power should be decreased.
QRQ	Shall I send faster?
	Sent faster (… in wpm).
QRS	Shall I send slowly?
	Sent slowly (in wpm).
QRT	Shall I stop sending?
	Stop sending.
QRU	Do you have anything for me?
	I have nothing for you.
QRV	Are you ready?
	I am ready.
QRX	When will you call again?
	I will call you again at … (time in hours) on (frequency in kHz). Minutes are implied
QRZ	Who is calling me?
	You are being called by … on … (frequency in kHz).
QSB	Are my signals fading?
	Your signals are fading.
QSK	Can you hear me between your signals? Can I break in on your transmission?
	I can hear you between signals; break in on my transmission.
QSL	Can you acknowledge receipt (of a message that's transmitted)?
	I am acknowledging receipt.
QSO	Will you relay to …
	I will relay to …
QST	General call preceding a message that is meant for amateurs and ARRL members.
	This is the "CQ ARRL" effect.
QSX	Will you listen to … on … (frequency in kHz)?
	I am listening to … on … (frequency in kHz).
QSY	Shall I change the transmission to another frequency (or on … kHz)?
	Change the transmission to another frequency (or on … kHz).
QTC	How many messages have you sent?
	I have sent … messages for you or for …
QTH	What's your location?

	My location is ...
QTR	What's the correct time? The time is ...

In addition to these abbreviations, there are also phonetic alphabets that are used by hams. These alphabets are used as standard while communicating on the air.

Letter	Word	Pronunciation
A	Alfa	AL FAH
B	Bravo	BRAH VOH
C	Charlie	CHAR LEE
D	Delta	DELL TAH
E	Echo	ECK OH
F	Foxtrot	FOKS TROT
G	Golf	GOLF
H	Hotel	HOH TELL
I	India	IN DELL AH
J	Juliet	JEW LEE ETT
K	Kilo	KEY LOH
L	Lima	LEE MAH
M	Mike	MIKE
N	November	NO VEM BER
O	Oscar	OSS CAH
P	Papa	PAH PAH
Q	Quebec	KEH BECK
R	Romeo	ROW ME OH
S	Sierra	SEE AIR RAH
T	Tango	TANG GO
U	Uniform	YOU NEE FORM
V	Victor	VIK TAH
W	Whiskey	WISS KEY
X	X-Ray	ECKS RAY
Y	Yankee	YANG KEY
Z	Zulu	ZOO LOO

Voice data: Voice is, by far, the most popular method of making contact. There are several ways of transmitting voice data using radio channels, and it is an easy and natural way to converse. Therefore, voice is used by hams for making short-

and long-distance contacts, and for public as well as emergency services. All those who have a Technician Class license can make voice contact with other hams, either directly or by using stations that relay transmissions from low-power mobile to a wider area (known as repeaters). There are also internet-connected repeater systems created by hams that are used to send digitized voices to different parts of the world. These systems enable the local repeater to communicate with others users across the world using only a low-power mobile or handheld radio.

To communicate within a country, hams generally use their local language; however, English is used as a common medium for making international contacts.

Digital data: With the advent of the internet and the emergence of advanced digital technologies, inexpensive signal-processing software and digitally-enabled hardware has become available. That has generated a growing interest in the digital modes of transmitting data over the airwaves. The data interface is used as a medium for connecting radio and computer. Most of the connections in this form are created by typing in words, using keyboards. Hams have also come up with new methods of converting signals into radio signals and vice-versa. The methods that are used for this conversion are known as protocols, for example: PACTOR, RTTY, etc. Different types of protocols are used for different radio communications, and the choice of protocols depends on the effect the transmission or reception has on the radio signal.

Exchanging emails over amateur radio looks very much like a normal email system and is used by many hams who are unable to use the internet due to connectivity issues faced while traveling to remote areas or being at sea.

Morse code: Morse code was quite a popular mode of communication in the old days of radio, however, even today it is commonly used in amateur radio. Since all the energy of a signal is focused in its single on and off, Morse code works efficiently with weak signals or in the presence of interference. Morse code can be generated using simple transmitters—devices that can create a signal and turn it on and off. A basic receiver is all that is needed for receiving the Morse code. Interestingly, Morse code is also used to create musical notes.

Novel Activities

Amateurs are known for developing new technologies in unique ways. Some examples include:

- **Packet radio** – With the advancement in the field of computer networks, amateurs have adapted various new protocols to operate over the air. For instance, they use a special interface that takes the characters from the computer, repackages them into data packets and transmits them using regular radio signals. The interface used is known as a Terminal Node Controller (TNC).
- **Automatic packet reporting system (APRS)** – This system was created to integrate GPS positioning and other data with packet radio. Amateurs with a mobile radio and access to GPS can send their location details to a local APRS gateway and even to servers that are internet-based. This can help other users find the location of the person who is sending the data by simply logging on to the servers and tracking the movement.
- **Amateur TV (ATV) and Slow-Scan TV (SSTV)** – ATV is used to broadcast TV-style videos by simply hooking up a normal video camera to an amateur transmitter. Once they do this, the videos are on the air and they look like a professional signal. SSTV was invented by hams to send images over normal voice radio signal. These images can be either colored or black and white, and can be received by a sound-card-enabled computer.

Chapter 2 - Getting Acquainted with the Technology

Before we dive deeper into the topic of ham radio, it is good to have an understanding of the underlying technology.

Understanding what makes ham radio work is difficult without having an idea of the purpose of radio, which is to send and receive data/information over radio waves—another form of light. Radio waves move at such a speed that they can traverse to the moon and bounce back within 2 ½ seconds; they can make a complete round of Earth in just 1/7 of a second. The energy of a radio wave is carried by both an electric and a magnetic field, which affect charged particles, such as electrons, in a wire. The charged particles move in a specific manner. For example, electrons move parallel and circular to electric fields due to the magnetic field effect. When these charged particles move, they further create moving electric and magnetic fields.

Transmitters bring electrons into motion, which, in turn, results in radio waves. When electrons move in structures like antennas, radio waves are created. These electrons also are impacted by radio waves created in other neighboring antennas. Receivers read this motion of electrons, created by inward radio waves, and the energy is transferred from electrons to radio waves and then back to electrons at the other station.

The relationship between radio waves and electrons is slightly strange. The energy strength of radio waves isn't the same all the time—it keeps varying between a positive and a negative value, and sometimes the variation is such that it looks like a vibrating string, moving up and down.

The amount of time a field takes to go through one complete set of values is known as a cycle. Hence, the *frequency* of a wave can be defined as the number of cycles completed by the wave in one second. It is measured in hertz (Hz). The amplitude or strength of a signal oscillates in a sine wave, and the period of the cycle is called its *duration*.

The shape of every cycle of the signal is basically the same—it rises and falls, and then returns to where it was. The position of a wave within a cycle is known as a *phase*, which is measured in degrees. One complete cycle of a sine wave is 360 degrees and if two sine waves have a 180-degree phase difference, which means one is decreasing while the other one is increasing, they are known to be "out of phase." When the two sine waves have no phase difference between them, meaning they're decreasing and increasing at the same time, they are known to be "in phase."

Another twist here is that the wave also travels at the speed of light, which is constant. But if you see the wave oscillating, you will notice that it always moves the same amount in one cycle, known as one *wavelength*. The higher its frequency, the less time it takes in completing a cycle. Waves that have a high frequency come with shorter wavelengths and vice versa; lower frequency waves have long wavelengths.

Therefore, if you know the frequency of a radio wave, you can easily derive its wavelength using the formula below (knowing the speed of light is constant):

Wavelength = Speed of light / Frequency of the wave

Similarly, if you know the wavelength of a wave, that is how far it traverses in one cycle, you can determine how fast it oscillates, as the speed of light is always constant.

Radio waves oscillate at frequencies between a few hundred kilohertz up to 100 gigahertz; hence, the corresponding wavelengths vary between hundreds of meters at low frequencies to a fraction of a millimeter at high frequencies. The two most common units used in measuring radio waves' wavelength and frequency are meters and megahertz.

Now that we understand the concept and know how wavelength, frequency and speed of light are related, let us dive deeper into the radio spectrum.

The spectrum of radio waves is so huge that when you tune your radio to these different frequencies, you can easily find different kinds of information being transmitted. These frequencies are also termed *signals* and are grouped together based on the type of information they carry in different frequencies, or in bands. For example, AM radio uses frequencies or bands between 550,000 and 1,700,000 Hertz.

The radio spectrum is also used by various kinds of services, such as ham radio services or broadcasting services. Each of these services is allocated a specific range of frequencies or bands to operate in, within the spectrum, which is known as a frequency allocation. There are a lot of frequencies allocated to the ham radio in the radio spectrum.

Radio waves of different frequencies behave differently in the way they travel and use other frequencies to receive and transmit. Since waves of similar frequencies have similar properties, the entire radio spectrum is divided into four sections—high frequency, VHF, UHF and microwave. Let's look at each of these segments in detail.

- **High frequency:** Frequencies below 30MHz are part of this segment. This includes the citizens band, AM broadcasting, some ham radio bands (10), military and ship-to-shore and ship-to-ship band.
- **Very high frequency:** Frequencies in the range of 30 to 300 MHz fall in this segment. This includes three ham bands, military, TV channels 2-13, FM broadcasting, cell phones and public safety.
- **Ultra-high frequency:** Frequencies in the range of 300 MHz to 1 GHz fall in this band. This includes two ham bands, military, TV channels 14 and above, cell phones, public safety and commercial mobile radio
- **Microwave:** Frequencies above 1 GHz fall in this band. This includes GPS, military, satellite TV, digital wireless telephones, microwave ovens, eight ham bands, Wi-Fi wireless networking and various public and private users

Modulation and its Types

Now that we understand the basic concepts of radio spectrum and waves, another important term we must discuss here is modulation. A simple radio signal, sitting quietly on the spectrum, doing nothing, isn't of much use. To be able to communicate, the data must be added on to the radio signal. This can be achieved using modulation. A simple radio signal that stands at one frequency, maintaining the same strength, is known as a continuous wave. When information is added to this signal by somehow modifying it, it is known as modulation. Modulation helps us to communicate information through radio signals. A signal that has not been modified, which means that it doesn't carry any information, is known as unmodulated. The process of recovering the information from a modulated radio signal is known as demodulation. To understand various techniques that hams use to communicate, it is important to understand the process of modulation.

There are different types of modulation, the simplest type being a continuous wave, modulated on and off in a coded pattern. For example, when Morse code radio signals, which are a continuous wave, are turned on and off, they are known to be modulated. If the information that is added onto the radio signal is 'speech,' the resulting signal is voice mode or phone. If data is used to modulate a signal, the resulting signal is a digital mode or data mode. The information carried by analog mode is understandable to humans, for instance, speech. Conversely, digital or data mode carry information that can be decoded only by computers or other machines.

The attributes of a signal that can be modulated are its frequency, strength and phase. All three of these attributes are modulated the in case of ham radio. AM and FM are the signals that we generally receive on our radio stations. AM stands for Amplitude Modulated signals and FM is Frequency Modulated signals.

Amplitude Modulation: Amplitude modulation, or AM, is the process by which the amplitude or strength of a signal is varied to add information (speech or data). The simplest form of AM is Morse code. If you want to see the strength of a signal being modified, place a meter in the path of your response signal or music. You will see that the amplitude of the AM signal changes to carry this speech or music data. An AM transmitter works in a very similar way—it adds your voice signals to a continuous wave by varying its strength or amplitude and the information carried is stored in the envelope of the resulting/modulated signal. On the other side, the receiver recovers this data signal or information from the modulated signal by following the variations in the signal's amplitude. This process of recovering the information that is carried by the signal, by following the envelope of the AM signal, is known as detection. Receivers are simple devices made out of galena crystal, which is a thin steel sliver, that perform detection in a simple manner. This is the reason for the popularity of the AM signal—it's easy to transmit and receive.

If we look at the AM signal carefully, it is made up of three components—a carrier and two sidebands. The carrier is a continuous wave with fixed strength and no information. The information is carried in the two sidebands. The total energy carried by the signal is divided among the carrier and the two sidebands.

Unlike the continuous signal that comes with uniform strength, there are composite signals. Composite signals are comprised of groups of signals that are combined together to create a single signal. The signals that are combined together are known as components and they cover a range of frequencies. The difference between the frequency of signal with lowest strength (lowest component) and the one with the highest strength (highest component) is called the bandwidth of that composite signal. In reality, continuous wave signals occupy the least bandwidth as compared to AM and FM.

If we look at the AM signal that is modulated by a single tone, it is comprised of two sidebands that are steady signals. The lower sideband has a frequency lower than that of the carrier wave and the upper sideband is higher in frequency than the carrier signal (this difference in the frequency depends on the tone). The information of the tone is stored in the amplitude of sidebands and the difference in frequency from the carrier. This can be better understood with the help of an example. Let's say an AM signal is modulated with a carrier of 800 kHz by a single steady tone of 0.6 kHz. When the receiver tunes to the AM signal from the left, it receives the lower sideband first, which is a steady signal at 799.4 kHz (800-0.6). After receiving the lower sideband, it receives the carrier at 800 kHz (which is 0.6 kHz higher in frequency). Lastly, it will encounter the upper sideband at 800.6 kHz (800 + 0.6). This way the three components combine to make an AM signal, where both the sidebands contain the information required to reproduce the tone that is used to modulate the carrier signal.

Since the process of modulation adds sidebands to the signal, the resulting signal is spread over a range of frequencies, known as bandwidth. For instance, the bandwidth of the resulting modulated signal in our example is 800.6 - 799.4 = 1.2 kHz. In this way, every modulated signal has some bandwidth, including the simple continuous waveform (CW) signal (bandwidth of around 100–150 Hz).

However, since voice, or any other data signal, contains several frequencies, and even the amplitude is constantly changing, the resulting modulated signal looks quite complicated, although the carrier is a continuous wave signal of constant strength.

If an AM signal is considered in terms of power, it is not that efficient, mainly because the carrier signal doesn't carry any type of information, although it uses a good amount of power. Furthermore, each sideband is comprised of a copy of the modulating signal. Hence, in reality, a fraction of signal is all that is needed, and that's what the single sidebands, or SSB signals, actually are. In SSBs, therefore, all power is accommodated in the one band that's available. Although these SSBs require complex equipment to be generated and received, their efficiency is so brilliant that creating them is worth all the trouble. This is because all the power is accumulated in a single band that carries the information. Another positive thing is that, since these SSB signals have smaller bandwidth (less than 3 kHz), they utilize the radio spectrum more efficiently. More SSB signals can fit in the allocated band or range of frequencies, without interfering with each other.

Frequency modulation and phase modulation: In the same manner that AM is created by modulating the amplitude, the other two modulations are created by varying the frequency and phase, respectively. When the frequency of a signal is varied to add data or speech information, it is known as Frequency Modulation or FM. In this mode, the frequency of the modulated signal varies with the amplitude of the modulating signal. When the phase of the signal is varied in order to add data or speech information, it is known as Phase Modulation or PM. In this mode, the phase of the modulated signal varies with the amplitude of the modulating signal. Unlike AM signal, FM and PM signals are comprised of one carrier and several sidebands that add up together. The resulting modulated signal has a varying frequency or phase, whereas its amplitude remains constant; the power also doesn't change, whether the signal is modulated or not. The amount by which the frequency varies, in the case of FM, is termed as deviation. For instance, when you speak louder in a FM transmitter, the deviation increases because the frequency is modulated with the amplitude. And when the deviation increases, the difference in the higher and lower band frequency also increases, increasing the bandwidth of the signal. Therefore, if there is too much deviation from the standard, it leads to interference with the nearby frequencies.

FM and PM are quite similar to each other and can be demodulated using the same receivers. In the case of ham radio, hams can use any frequency within the given band, but they must operate carefully while using the frequency that's near the edge of the band. That's because they have to use the frequencies within the band and, as the radio displays the carrier frequency, you must also always leave room for the sidebands. For instance, if the frequency modulated signal is 12 kHz wide, the carrier frequency should be in the center of the signal, and therefore should never be less than 6 kHz from the edge of the band.

Now that we know the different types of modulation available, let us compare and see which type to use. This requires understanding the different types of signals and modulations available, their strengths, shortcomings etc.

FM occupies more bandwidth than the SSBs. This is because the FM signal is created by varying just the frequency. That's why all electrical and atmospheric noises that are received as part of amplitude variations do not interfere with the FM signal. The FM receiver is comprised of a limiter that strips away all such amplitude variations in the incoming FM signal, enclosing these noises, and that's why they do not cause any disturbance in the output signal—it all gets filtered out. All these noises impact the AM signal, which is why programs broadcast on AM frequencies often experience noise crashes and disturbances, whereas programs broadcast on FM stations are usually disturbance-free. Though they afford such great benefits, bandwidth considerations are often overlooked. Frequency modulation is also used for data signals for VHF and UHF. With the help of FM radio speech input and audio output, the data is transferred as audio tones. Thus, FM voice radios are used.

On the other hand, SSB is used in cases where signals are comparatively weaker and spectrum space is not enough to support the huge bandwidth of FM signals. Since the power of an SSB signal is accumulated in a narrow bandwidth, unlike the FM signals, they can be used for communication purposes over longer ranges. They work fine even in poor conditions when FM and AM fail to work. That's the reason VHF and UHF contest operators and DXers use SSB signals. If you are looking for even better results, CW signals (that have bandwidth narrower than SSBs) are the easiest ones to send and receive, especially if conditions are poor or fading. Hams refer to SSB and CW as weak-signal modes as they are quite effective at low-signal strengths, unlike FM and AM.

Another important thing we must know is when to use which sideband (as SSBs can use either use the upper sideband or the lower one). Although there is no technological reason involved here, hams have standardized the following conventions to improve communication:

- LSB should be used if frequency is below 10 MHz.
- USB should be used for all frequencies above 10 MHz.

Basics of Electricity

After understanding the basics of waves, let's have a look at the fundamental principles of electricity. Knowing the basic concepts will help us in understanding the sophisticated electronics that are used in ham radios.

The moment we think about electricity, the first term that comes to most of our minds is current, which is the flow of electrons. Electrons are negatively charged, whereas current is measured when these electrons flow through a carrier, such as a wire. It is measured in amperes, or amps, using an ammeter. Electrons are tiny particles; they are so tiny that to create a current of 1 ampere, billions of electrons are required to pass through the device each second.

Another commonly used term is voltage, which is the electric potential or force responsible for the movement of electrons. It is always measured with respect to a reference, or from one point to another. This is measured using a voltmeter. The unit of voltage is volts. Unlike electrons, that carry a negative charge, voltage can be positive (attracting electrons) or negative (repelling electrons). The surface of the earth is considered to be the reference for measuring voltage and is known as ground potential.

In the same way that the laws of electricity work with negative charges flowing backward, they work with positive charges flowing forward. When positive charges move from a positive to a negative side, this is known as a conventional current, and its mirror image is known as an electronic current.

When a current completes its path, this is known as a circuit. When multiple devices are connected in a circuit in a way that the same current flows through them, this is known as a series circuit. On the other hand, if multiple devices are connected in a way that they have same voltage across them, it is known as a parallel circuit. If voltage is to be measured, the voltmeters will have to be connected in parallel with the given component. And when current is to be measured, the ammeter is connected in series with the given component so that the same current flows through it.

All these are basic test instruments—ammeters, voltmeters and ohmmeters. But we cannot have a different instrument for each parameter, and that's why experts invented a multifunctional meter, known as a multimeter. A multimeter uses a single meter to measure all the three basic units—voltage, resistance and current—and is also known as a volt-ohm-meter or VOM. Multimeters can be analog or digital and their functionality and input depends on their type. For example—a typical digital multimeter comes with a switch and set of inputs to select which ranges of values and parameter are to be measured. Although their functionality is a little complicated, as compared to single meters, the advantage of having just one meter to measure all the parameters overshadows this added complexity. Sometimes there are strange ways of interpreting their readings. For example, if you are trying to measure the resistance of a component and the reading initially is very low, but increases with time, it means the circuit is comprised of a large value capacitor. Another thing to always keep in mind while using multimeters is that they should be used carefully and properly. If you try to measure voltage or current in a circuit using the multimeter that is set to measure the resistance, you can damage it. You should also take into consideration the maximum value it can measure, because if you try to measure a value above its capacity, it might overflow to other equipment in its surroundings, or to you, which can be really dangerous. So ensure you measure the right parameter and right values.

Just like voltage and current, resistance is another important parameter in the world of electronics. We know that the current is caused when the electrons flow through a component. However, this flow is always restricted, to an extent, by certain type of materials. This property of impeding the flow of a current is known as resistance, which is measured in ohms, with the help of ohmmeter. There are certain types of materials that offer minimal resistance to the flow of electrons in the presence of applied voltages. Such materials are known as good conductors of electricity, for example—copper. Then there are materials that prevent the flow of electricity or offer high levels of resistance to the flow of electrons. These are known as bad conductors of electricity or insulators, for example—paper, ceramics, glass, etc.

Scientists in the nineteenth century discovered that current, voltage and resistance are proportional. To be precise, current is directly proportional to voltage and inversely proportional to resistance. If the resistance in a component is high, there will be less flow of electrons and hence, a low current. On the other hand, if there is a high voltage applied across a circuit, there will be more flow of electrons and hence, a lower current.

Another important parameter is power. Power is the rate at which energy is utilized. The value of this parameter can easily be derived by multiplying voltage and the current. This means,

P = E x I, where P is the power, E is the voltage and I is the current

Power is measured in watts.

Now that we are aware of all the key parameters used in the world of electronics, let us look at the basic components used. Electronic circuits are formed when various types of components are connected with each other in a specific manner. The way they are connected has an underlying logic. Each component of a circuit plays a key role—some of them help in routing a current, some amplify a signal, while others dissipate the energy from the circuit. Looking at these different components can help us understand how the circuit works.

The most basic types of components include capacitors, resistors and inductors. All three components have a different impact on current and voltage and, thus, are measured differently. Note that the wires that are used to connect these components in a circuit have a minimal impact on their functioning.

Resistors: These components add a resistance in the circuit. Different resistors are available in the market, with different values of resistance, measured in ohms or kilo-ohms. They are added into the circuit whenever there is a need to oppose the flow of electrons, acting just like a valve in a pipe that is added whenever the flow of water has to be restricted. When a current passes through the resistor, some amount of energy is lost.

Capacitors: These are the components that store energy in the electric field that is created when voltage is applied between the two conductors, known as electrodes, that are separated by an insulator. So a capacitor is comprised of two electrodes and a separating, insulating dielectric between them. The stored energy in this component is known as capacitance and is measured in farads (picofarads, nanofarads and microfarads). Since they store and release energy between the two conductors, as needed, capacitors are normally used to smooth out the variations in voltage in a circuit. They cannot pass on the direct current (DC), since the two electrodes are separated; however, they can pass through the alternating current (AC).

Inductors: Just like capacitors, inductors store energy, but the difference is that they store it in the magnetic field that is created due to the flow of current through a conductor. An inductor is made out of a wire wound around the coil of magnetic material. The stored energy in this component is known as inductance and is measured in henrys (nanohenrys, microhenrys and millihenrys). Since they store and release current in the circuit, as needed, inductors are normally used to smooth out the variations in current in the circuit. Just like capacitors, inductors don't work for DC but resist the flow of AC.

Since all these components are available in different values, they are sometimes available in different colors so that they can be differentiated easily, based on the values they hold. For example, a resistor made with four colored strips—red, violet, brown and gold—is created to have a value of 270 ohm (each of these colors holds a different value). The last color, gold, signifies a tolerance of five percent.

All the basic components—capacitors, inductors, etc., can be variable or adjustable. The variable versions have their own specific uses. For example, variable inductors and capacitors are used to tune the radios. Then there are transformers, which are made using multiple inductors that share the energy stored. The energy flows from one inductor to another when the combination of voltage and current is changed.

AC currents are in phase, and AC voltages are in step in a resistor. When the voltage goes up, the current shoots up, and vice-versa. Similarly, in inductors and capacitors, to have an offset in time between changes in AC voltage and changes in AC current, the relationship between the two is changed, leading to the storage and release of energy. When this happens, a phase difference occurs between the voltage and the current. On the other hand, in a capacitor, current changes are ahead of voltage changes due to the fact that the smoothing action of the component works against the voltage changes. In the inductor, current changes lag behind voltage changes as the components oppose the current changes. This leads to hindrance in the path of the AC current, known as reactance. This reactance is measured in ohms.

Reactance created in a capacitor is known as capacitance reactance, whereas reactance created in an inductor is known as inductive reactance. The value of reactance of any component depends on the frequency of AC current and the capacitance or inductance, whatever is applicable. The combination of two—reactance and resistance—known as impedance, is measured in ohms. Since we get to see both reactance and resistance in radio circuits, impedance is often used in the world of radio.

Since the offset in time for voltage is the opposite to that of a current, the impact of capacitive reactance on an alternating current is different from that of inductive reactance. Therefore, in a circuit where there is capacitive reactance and inductive reactance, there will be a frequency at which the value of both will be equal, and hence, they will cancel each other out. At this frequency, the voltage and the AC current are back in step, giving rise to a condition known as resonance. The frequency at which resonance occurs is known as a resonant frequency. All the circuits that comprise both inductors and capacitors will have at least one frequency where the impact of two reactance on AC current will be negated. When this happens, the circuits are known to be in resonance, or tuned. Therefore, a resonant or tuned circuit is often used as a filter to reject or pass signals at the resonating frequency. And these circuits play a crucial role in the life of radios as they help them generate or reject signals at a particular frequency.

Advanced Electronic Components

We know have a basic idea about the most commonly used components—capacitors, inductors and resistors. All three components treat the polarities of current and voltage as the same. Then there are other types of components that react based on the polarity of a current and voltage, and these are generally created out of semi-conductor materials. Semi-conductor materials are neither good conductors of electricity nor are they good insulators. However, there are certain types of semi-conductors, for example, silicon, that have a unique ability to change their behavior upon adding little bit of impurities, known as doping. Depending on the nature of impurities added, the material shows N-type or P-type of behavior. An amazing thing about these N-type and P-type materials is that, when they are placed in contact with each other, they create a PN junction, which is a better conductor of electricity in one direction as compared to another. This property is often used to create some of useful components.

There are certain semi-conductors that allow the current to flow only in one specific direction. Such semi-conductors are known as diodes. A diode is comprised of two electrodes—an anode and a cathode. Diodes that are heavy-duty, and can handle large amounts of current and voltages, are known as rectifiers. When an AC voltage is applied to a diode, it gives rise to unidirectional DC current. This is because when voltage is applied to a diode that can let the current flow only in one direction, the current is hindered because voltage tries to move the electrons in the wrong direction.

Then we have an advanced type of diode, which emits light, and hence is known as a light-emitting diode or an LED. When current flows through this diode, it gives you energy in the form of light, the color of which is determined by the material that it is made up of. LEDs generally require less power to function and hence are used as an effective alternative for incandescent lamps.

Other components that are made out of semiconductors are transistors. These are created from the N-type and P-type material, when they are arranged in a manner that lets it use small currents and voltages to control the bigger ones. There are two types of transistors—bipolar junction and field-effect. The bipolar junction transistor is created by arranging three alternating layers of N-type and P-type material, known as base, emitter and collector. These types of transistors can be categorized under two heads—NPN and PNP type, depending on the arrangement of the three N-type and P-type layers. On the other hand, the field effect transistor is created as a channel of N-type and P-type materials. The two ends of this channel are known as the source electrode and the drain electrode. Then there is a gate electrode, which manages the flow of the current through this channel.

These are the components that are often used in the electrical appliances. Based on the requirements and needs of an appliance, various components are connected with each other and packaged as one unit, which is known as an integrated circuit, or IC. Integrated circuits are the package set of components that can range from simple circuits, comprised of only two or three diodes, to some really complicated circuits that combine hundreds or thousands of components.

There is another category of components, known as protective components, which are used to prevent damage to a circuit, such as fire or a shock. These components are designed so that they do not have any impact on the working of the circuit under normal circumstances. When there is a malfunction, protective components are triggered and prevent damage from happening. Therefore, it is quite important to understand the various types of protective components that are available today, to see how best we can use them to protect our valuable appliances.

A fuse is a type of protective component that is triggered whenever there is a current overload. It interrupts the current flow by melting away some amount of metal, which breaks the circuit. However, once used, the fuse cannot be reused. Fuses are generally rated in terms of the maximum amount of current they can handle without blowing off the circuit. There are also slow-blow fuses that are capable of temporarily handling overloads but, if exposed to consistent overload, they blow off. Then we have circuit breakers, which are just like fuses as they break the circuit whenever there is a current overload. The difference is that fuses melt away metal, whereas circuit breaks trip a circuit, and therefore can be reused by resetting, unlike fuses.

Another type of protective component often used in AC power wiring is a ground fault circuit interrupter. It trips the moment it senses an imbalance in the current that is being carried by the conductors. Current imbalances are really hazardous as they can lead to electric shocks.

Whenever there is a need to change a circuit breaker or a fuse, one must always replace it with one that has the same current rating, to avoid any kind of damages or hazards. Using a protective component of a different value, even for a short amount of time, can lead to permanent damage of the component or entire circuit.

There is a type of protective component known as a surge protector. This protects devices from high voltage by hindering the temporary transients in the voltage that fall outside the normal range by turning into resistors. Then the energy that would have otherwise been passed on to the device is dissipated in the form of heat energy. Surge protectors are generally used in residential AC circuits, telephone lines, etc. Another protective component that works along the similar lines is the lightning arrestor, which is used to handle much higher voltages.

Another special type of component is a circuit gatekeeper. Circuit gatekeepers are so named as they control the amount of current that flows in the circuit. Relays and switches are simple types of circuit gatekeepers that control the amount of current that passes through a circuit by disconnecting and connecting the path it can follow. When the circuit is opened, the current is interrupted, and when it is closed (meaning complete), it flows again. While a switch is manually operated, a relay is operated using a switch with the help of electromagnetic energy. There are different types of switches and relays available in the market and they are differentiated based on the number of throws and poles they have. Each pole is responsible for controlling the flow path of a current. For example, a double-pole switch controls two different currents. Each throw refers to a unique path of the current. For example—a double-throw switch can make the current flow in either of the two paths but a single-throw switch can only make it flow/not flow on a single path. The two concepts, combined together, explain how a switch functions. For example, a single-pole single throw, the simplest known switch, closes or opens a single-current path.

Other important components that are often used in radio equipment include displays, meters and indicators. An indicator is used to show either an on or off state. Radio equipment often uses various indicators. A meter, on the other hand, is used to display value in the form of numbers. When the two—indicator and meter—are connected together, they become a display. For example—a liquid crystal display, or LSD, is commonly seen on the panels of a lot of equipment.

Understanding Basic Radio Gadgetry

Although we still might see vacuum-tube radios enhancing the beauty of an antique-loving ham's radio station, most radios today are sleek and controlled by microprocessors. The basic framework is comprised of a receiver and a transmitter—a transreceiver that is known as a rig by hams. Handheld radios and mobiles are also known as rigs. If these so-called rigs do not operate on AC line power, then there is a need to use a power supply to give them DC current and voltage.

After from a transmitter and a receiver, the third basic element of a radio station is the antenna, which transforms signals from a transmitter into energy that is capable of traveling through space as a radio wave. This antenna also performs the job of capturing these radio waves and converting them into signals that can then be received by the receiver. Most transreceivers share a single antenna for transmitting and receiving the signals by using a transmit-receive (TR) switch. What connects the antenna to the transmitter and receiver is a feed line, also known as a transmission line. Feed lines are just like the power lines that are used to transfer energy; only feed lines transfer radio signals. Hams call these antennas skyhooks.

There are two types of antennas that are usually used—dipole and beam antenna. A dipole is the antenna that is comprised of wire and connected to the feed line in the center. Conversely, a beam antenna is efficient in sending and receiving radio waves in one direction; thus it is mounted on a tower with a rotator that points in different directions. For the transmitter to be able to operate at peak efficiency, an antenna tuner is added between the transmitter and the antenna/feed line.

A station that is comprised of several antennas has an aluminum farm. Terms such as aerial and set are no longer used by hams, as they are considered obsolete.

Apart from all these components, there are certain miscellaneous gadgets that form part of the actual operating system. Some of these include:

- **Feed-line measurements:** Most antenna tuners and radios are capable of evaluating the conditions inside the feed line, measured as standing-wave ratio, or SWR, which is a ratio of voltages. SWR determines how much power, supplied by the transmitter, is radiated by the antenna. The majority of radios are equipped with an inbuilt meter to measure this indicator, but some choose to use a stand-alone SWR sensor, called an SWR meter. Feedline conditions can also be measured using a power meter, which measures the amount of power flowing through. Most buy

SWRs meter, as power meters are more expensive, although they are quite accurate in measuring the conditions.

- **Filters:** These filters are designed to pass and reject a range of frequencies. They can be made out of capacitors and inductors, which are termed as discrete components, or from sections of feed line that are called stubs, or can contain amplifiers. Below are some types of filters that are commonly used:
    - Receiving filters – These are installed inside the radio circuit and are normally made from small tuning-fork-like structures or quartz crystals, which are so designed that they let only certain frequencies pass, and block others. They are also known to improve the selecting of receivers, which means they are capable of receiving one signal when there are multiple signals present.
    - Feed line – Feed-line filters are designed to prevent undesired signals from the antenna to the radio, or vice-versa. When used on transmitted signals, they are used to make sure the unwanted signals are not radiated out from the transmitter, causing a disturbance or interference in the desired frequencies. When used on the receiver side, they prevent the receiver from receiving signals that are not desired, thus improving the performance of the receiver.
    - Notch – These filters work towards removing a narrow range of frequencies, like a single interfering tone.
    - Audio – Audio filters are used to enhance the filtering capability on the receiver end, thereby rejecting any unwanted signals in the surroundings.
- **Feed lines:** Antennas are connected to the station and carry energy between the equipment with the help of two types of feed lines: coax cable and open wire. A coax cable is comprised of a hollow tube that surrounds a central wire in a concentric manner. The outer layer is called the shield, and if it is made from fine woven wire, it is called a braid. The wire in the center is called the center conductor, and to hold it tight in place, it is surrounded by insulation. The outer conductor has a covering made up of plastic, known as the jacket.

Open wire is another type of feed line made from parallel wires, which is also known as a ladder line. The insulating spacers hold them together and may have plastic insulation to shield them.

- **Microphone:** A microphone converts the voice of an operator into an electrical audio signal.
- **Oscillators and amplifiers:** Oscillators produce a constant value signal at a given frequency. These are used in both transmitters and receivers of radio signals to describe the frequency that is being used. Before transmitting, the output signal of the oscillator is modulated, and then amplified, before being turned over to the antenna for transmission. The amplifier is the circuit that enhances the strength of the signal so that it doesn't die during the transmission. It is needed for reliable communication over long distances.
- **Modulators:** We discussed the process of modulation in detail earlier in this chapter. The process of combining a signal carrying information with an RF signal is known as modulation. The circuit that performs this task of modulation is known as a modulator. In simple words, a modulator is a circuit that adds the data or information signal to the carrier signal, resulting in an RF signal that can be transmitted over a radio frequency. Contrary to what the modulator does, a demodulator circuit works towards extracting the information from the modulated RF signal.
- **Mixers:** A mixer is a close friend of a modulator. Both circuits combine two signals to create an output signal of a different frequency. The mixer combines two RF signals in a way that the resultant signal is at a different frequency. They are used in both transmitters, as well as receivers, to move one of the signals to a third frequency for specific purposes.

Chapter 3 – Handle Your Equipment Well

If running a ham radio station required engineering/technological or programming skills, very few would dare to enter the arena! At the same time, you cannot be a complete ignoramus and expect your equipment to take care of itself! After all, you are dealing with both electricity and electronics. Therefore, you need to have some very basic skills, along with relevant knowledge at your fingertips, in order to keep everything functioning smoothly. Yes, there will be initial hitches and glitches, since you are a novice in this field. However, as you move ahead with confidence and courage, you will find the problems shrinking, until they almost disappear. You will also be able to go on the air more frequently than before, enabling you to improve your contact base. Apart from this, you will not hesitate to experiment with novel bands/modes, in the hope of enhancing your education and talents. Thus, if you protect your radio against breakages/damages, understand what has gone wrong at certain times, and understand how to repair broken/damaged parts with the right tools, you should be able to perform your operations smoothly.

A. Basic Test Kit

While it is good to obtain all the required tools to keep your ham radio station in good running order, it is equally important to have a basic test kit in place, comprised of instruments that will help you operate the gear lying around your station. You are free to choose what you would like to keep in your kit. These are some suggestions.

DVM/VOM (Digital Volt Meter/Volt-Ohm Meter)

This is an extremely handy gadget for monitoring voltages, resistance, continuity of electric flow, current with suitable shunt circuitry and so on. Do not skimp on this instrument, for you will need to use the multimeter very frequently. Get the best one that will fit into your budget, after taking advice from experienced professionals. Pricing varies vastly, from as little as $5 to as high as $200 and beyond! Take note that the probes accompanying the instrument have long handles. It will not do for your fingers to encounter the circuit which you are testing! There should be a flexible and strong wire in place. This wire remains attached to the instrument at one end. The connection between the test probe and the plug should be tight and strong. If you have to keep soldering this connection frequently, you have wasted money on an inappropriately functioning instrument. If possible, consult professionals, well-written handbooks, etc., for guidance on how to build simple devices, which will help you enhance the capabilities of your VOM/DVM.

VSWR (Voltage Standing Wave Ratio)

This is the power meter. Models range from highly basic and affordable to extremely complex and costly. Some instruments display the reflected power, or SWR, at the mere push of a button. Others possess two meters, placed on a single dial, for simultaneously displaying both the reflected power and the forward power. This is the cross-needle display. A third variety digitally shows the SWR and/or power. Note that the degree of accuracy varies from specimen to specimen. Therefore, you will have to be careful by approaching a genuine dealer for an authentic instrument. Furthermore, obtain something that offers maximum accuracy.

Antenna Analyzers

These instruments provide highly accurate and relevant information, if they are of high quality. Yet they tend to be rather expensive. Therefore, do some research to see if you can operate without them.

Sometimes, a particular SWR meter feels left out in the line. This means that when you connect the SWR between the antenna and the transmitter, and key the mic, you will be able to read either the forward power or the standing- wave ratio. Since you need to measure the SWR, put the instrument back in line. Then, adjust the antenna, such that the reading becomes acceptable. Next, detach everything and plug the antenna directly into the tuner/radio, as you think best.

When a VSWR meter has call liberations for high frequencies, you will not obtain accurate measurements while utilizing it on 144 MHz. Similarly, a set for 144 MHz give you the SWR ratio of a 440 MHz antenna. Therefore, you will have to match your meter to a particular radio band.

Oscilloscope

Do you love to experiment with building all kinds of kits? Alternatively, would you love to learn more about your radio, as well as how it operates? If your answers are in the affirmative, you should look into buying an oscilloscope, which is available in a digital or analog format. There are even oscilloscope software programs. Of course, the price will vary, depending upon the quality. You can look for second-hand goods, provided they are of good quality, at hamfests. Ensure that you know what you are doing before opting to refurbishing a used oscilloscope. New pieces will prove even more expensive than high-quality HF radios.

Frequency Counter

This is quite handy for a ham radio operator. It will help you observe your transmitter's frequency. Is the gadget emitting the frequency that it should be emitting, or not? Price is dependent upon the degree of accuracy that each instrument displays. Even the lowest price may seem high for you, since you are just starting out and may have already spent a lot on acquiring necessary equipment. Therefore, you have to decide if you truly want this nice-to-have instrument or not. After all, you will not be using it every day.

Signal Generator

This instrument is great for figuring out the accuracy of your receiver. At the same time, you may not require it much if you are not planning to engage in a lot of construction or repair work. Therefore, consider the cost/benefit ratio, prior to an actual purchase.

Spectrum Analyzer

This is similar to an oscilloscope, for it provides a visual display of an electrical signal. The difference is that the spectrum analyzer measures the signal against frequency, while the oscilloscope measures it against time.

You will require some training in order to handle the spectrum analyzer appropriately. The purchase and the training may be expensive. Do you really want to take such a risk in the initial stages? Therefore, it might be well worth it to leave the instrument to the professionals, as well as to the few ham operators working from home.

In conclusion, you would do well to begin with a DVM/VOM and VSWR in place. You do not really need the other instruments at the start. The solution to any problem might be quite simple, if only you keep a cool head! It should not prove too difficult to consider a fuse, coax, antenna, connections, etc., before you panic and order repairs or replacements for diverse types of gear!

B. Requisites for Building and Repairing

As mentioned earlier, you will need some basic tools in place, if you have to manage all the equipment in your ham radio station. In fact, you will have to pamper everything, in order to keep things running smoothly. At the same time, don't just buy every piece of expensive equipment that's recommended, just because you feel that you may need it in the future. Just obtain what you need in order to improve your knowledge of electronics and electricity via practical handiwork.

Shopping Tips

It is all too easy to allow unscrupulous dealers to take you for a ride! In fact, you may end up purchasing things that you do not really need and leave out essential components for your ham radio station. Here are a few recommendations for shopping with a sensible head on your shoulders.

- Let hamfests be your first choice, always, and not local shops or dealers. You should be able to find all manner of bargains in components at these trade fairs. Furthermore, there are experienced people around who will willingly give you advice. You should be able to find switches, complex goods, drawers/cabinets for holding your components, etc., at such places.
- Whenever you order or purchase components for any project, buy extras. In fact, you will save money when you buy smaller items (transistors, resistors, diodes, capacitors, etc.) in bulk. Any dealer will reduce the price if you request amounts of ten or more. Soon enough, you should have a nice collection for undertaking newer projects!
- You may reduce your shopping trips if you take note of all the entertainment and appliances lying around the house. Several of them may be broken or damaged. Prior to throwing these devices or appliances into the trash, strip each one of all the materials that may prove useful for your ham radio station. For instance, you will find speaker jacks, power

cords, switches, transformers, headphones and so on. You will even receive lessons in how all this hardware came into being, while stripping the appliances/devices!

- Have a hardware junk box in place. It can be a flat and open tray or an old paint tray. A real box will make rooting for things all that much harder. Now, whenever you find loose nuts, screws, springs, spacers, etc., toss them into the junk box. Think of all the time you will save when you need some essentials in a hurry!

Tools to Build and Repair

Outlined below are some tools that should come in handy for tackling a variety of construction and repair projects. They are the commonest of components, which should exist in every ham radio shack.

ICs (Integrated Circuits)

They include audio amplifiers, voltage regulators and so on. There are loads of them on the marketplace. However, what you really need at your shack includes an LM386 audio amplifier, a MAX232 RS232 interface and an LM555 timer. You may also benefit by having these voltage regulators positioned on your shelf—7805, 7812, 7912, 78L05 and 78L12. Keep op-amps LM358, LM324 and LM741 around too.

Capacitors

These devices store electric charges and are comprised of a lone pair, or more pairs of conductors. These conductors remain separate from one another with the help of insulators. Since you are dealing with electrical and electronic equipment, you should find them handy. Obtain ceramic (0.001, 0.01 and 0.1 µF), electrolytic (1,000 and 10,000 µF) and tantalum/electrolytic (1, 10 and 100 µF). You may also go for other values between 220- and 10,000-pF ceramic/film.

Resistors

They show resistance to electric current flowing through a passage. Go for values ohm-variable resistors and controls (100, 500, 1k, 5k, 10k and 100k), diverse values of five percent carbon/metal film and fixed-value resistors (1/4- and ½-watt).

Inductors

These are components in electronic or electric circuits, which produce impedance. Buy chokes of 1, 10 and 100 mH.

Semi-Conductors

These solid substances exhibit conductivity between a majority of metals and an insulator. This is because some impurities have gotten in the way, or because of the effects of temperature. Generally, silicon semiconductors come into play in electronic circuits. Therefore, you should obtain red- and green-colored LEDs. Along with them, buy full-wave bridge rectifiers, as well as 1N4001, 1N4007 and 1N4148. Then again, there are switching transistors (2N2222, 2N3906 and 2N3904), as well as IRF510 FETs.

Specialty Tools

Since you work with both plastic and metal substances, you will need certain instruments and tools for taking measurements and making requisite adjustments.

T-Handled Reamer and Countersink

Whenever you need to make a small hole bigger, choose the T-handled reamer. It will create a precise fit. Similarly, a countersink helps to smooth out the edges of the hole that you have made. It also removes all the burrs.

Nibbling Tool and Chassis Punch

A nibbler is a punch that you may operate manually for biting off small chunks of plastic or metal. The sheets possess a rectangular shape. Of course, the hole will not be regular in the beginning. You will have to use a file to shape it. In contrast, the chassis punch, which is used on steel or aluminum sheets, creates perfectly clean holes. The sheets may go up to 20-guage steel or 1/8-inch aluminum. Neither instrument is cheap. However, if you are planning to be a regular builder, then you will save enormously on time and quality by obtaining good instruments.

Ordinarily, you would not need to adopt special measures for creating a hole in a chassis or panel. However, if the chassis or panel displays electronics or wiring mounted on it, or even near it, you will have to undertake special precautions. You cannot allow metal chips to drop into the equipment. Then again, you must prevent the drill from going in too deep. You can protect the side of the panel where the drilling is going to take place by placing a few layers of masking tape on it. Ensure that the outer layers of your 'mask' remain slightly loose, in order to take the place of a safety net. Similarly, take a small piece of hollow tubing and place it on the drill bit. Place it in such a way that only a part of the drill bit is exposed. This exposed part should suffice to indicate how far the bit should penetrate into the panel or chassis.

Wattmeter or SWR Meter

Considering that you are dealing with electric and electronic equipment, these are bound to experience malfunctions or wear and tear. Therefore, you may require a separate device for measuring the power of your transmitter. You should be able to purchase inexpensive models, provided you know how to recognize quality, at diverse shops. Avoid going to CB shops, since these instruments do not display proper calibrations. Instead, opt for the Bird Model 43, which ham radio enthusiasts believe to be the gold standard. As far as diverse frequencies and power levels are concerned, slugs or varied elements come into play. You should be able to find both used elements and meters in the marketplace. Always keep quality in mind as you're shopping.

Oscilloscope, Audio and RF Generators

Since radio deals mainly with alternate current signals, you must know how to generate them and view them. You will be able to gain better guidance about purchasing them by perusing reputed websites.

C. Tools for Maintenance

There are two sides to maintenance. One is to care for all your existing equipment. The other is to fabricate whatever fixtures or cables you require at that moment, and assemble them into a complete whole. You require a good set of tools on hand to undertake both these tasks, specifically if electronic equipment is involved.

VOM (Volt-Ohmmeter)

You will need to be able to afford this instrument, for it does not come cheap. In case you can purchase it, opt for something that checks both transistor and diode. It should also behave as a continuity tester. It is possible to add an inductance checker and capacitor to the instrument. Certain models offer a frequency counter too, which should prove extremely handy.

Soldering Iron and Gun

Choose a model that offers adjustable temperature, as well as interchangeable tips. After all, printed circuit boards and fragile connectors cannot function in the absence of a low-temperature and fine-point tip. If you intend to tackle heavier wiring jobs, then you will need something with a bigger tip. It should also provide more heat while working. Therefore, obtain a soldering gun that has at least 100 watts of power, perfectly suitable for soldering jobs involving cables and antennas. Do not use a soldering gun for tackling smaller jobs.

Head-Mounted Magnifier

There is no denying that your eyes suffer immense strain while staring at tiny electronic components. In fact, these components are only getting smaller by the day! Therefore, you should buy a head-mounted magnifier. Craft stores should offer good quality instruments. Alternatively, you may opt for clamp-mounted, swing-arm magnifier-light combinations.

Terminal Crimpers

It is imperative that you comprehend how to install crimp terminals with the right set of tools, which are terminal crimpers. This is because you may not use ordinary pliers on crimp terminals. If you try to use them, the connection may pull out entirely, or merely work loose. As a result, you will have to spend hours figuring out how to bring the connections together again. You may require some help in using the terminal crimpers at first. However, there should be plenty of advisors around to help.

Wire Cutters

They should not be of the ordinary kind, but a heavy-duty pair. After all, you will need to tackle big wires and cables. Also keep a pair of diagonal cutters handy.

They should be extremely sharp and have pointed ends (dikes). The cutters will help you tackle smaller jobs with ease.

You will need to keep some cleaning equipment at the ready, too, for this is an important aspect of maintenance.

Solvents and Sprays

There are many brands on the marketplace. It is your choice to decide what you require for your gear. Whenever you wish to clean cabinets and panels, you might opt for lighter fluid. It is available in cans or bottles. The lighter fluid is useful because it allows you to perform 'gentle' cleaning. You should be able to do the job quickly, removing old tape and adhesive. With regard to other solvents and sprays, you might want to keep a bottle of isopropyl alcohol, along with compressed air and contact cleaner. Whenever you wish to use a solvent, test it on a small, hidden piece of plastic first. If it works, spread a larger quantity over your gear.

Soft-Bristle Brushes

You do not need to purchase brand-new brushes for your cleaning tasks. Discarded toothbrushes and old paintbrushes will be adequate. However, the paintbrushes should be small. Use a round brush for cleaning holes and tubes.

Metal-Bristle Brushes

If you want to clean up corrosion and oxide, choose light-duty brass or steel brushes. Note that while brass brushes have no negative effects on metal connectors, they will definitely damage plastic displays or knobs. Therefore, steel brushes are more useful in this scenario. Do not forget to clean your brushes after you have used them.

D. Spare Parts, Adapters and Common Consumables

Spare Parts

It is not enough to stock up on tools and leave it at that! You need to have spares handy too. Towards this end, obtain a spare part for every connector in the place. It is not necessary for every piece of your gear to function on the same connector. There are bound to be differences. Therefore, make a list, and obtain all the spares. If you manage to get a couple of spares for each connector, that's not a bad idea. They undergo wear and tear over time, and you will always have a need for the spares. With regard to coaxial cables, you will need some spare RF connectors. Go for the commonest types, which are N-Type, BNC and UHF.

Since the focus is on connectors, you should understand the differences between a jack, plug, socket, female connector and male connector.

- Jack refers to the connector, which you mount on a piece of equipment.
- A plug is the connection between the cable and the electric socket. It remains attached to the end of a cable.
- A socket refers to a receptacle, which allows an electric device to insert itself.
- A female connector is a recessed socket. You may insert a male connector pin into it.
- A male connector has exposed pins, which function as signal contacts. You need not consider the outer shell/shroud in this matter.

Adapters

Whenever you cannot find the perfect cable for your equipment, you will have to seek an adapter. It is the same when a novel accessory possesses a different type of connector. Towards this end, you may use a double-female (barrel) adapter for three types of connectors. The BNC plug adapts to the UHF Jack (SO-239). Similarly, the UHF plug is compatible with the BNC jack. Even the combinations of UHF jack and N-Type plug or the UHF plug and N-type jack work fine. These adapters are RF types.

Audio adapters are great for working on mono- to stereo-phone plugs, RCA double female for splices, etc. The data adapter is for the 9-pin to 25-pin D type, BNC plug/jack, N-Type plug/jack and UHF plug/jack, etc. Bear in mind that you do not need to purchase all manner of adapters at one go. However, whenever you go to a hamfest, obtain whatever you need at reasonable prices.

Common Consumables

They are as important as spare parts and adapters.

- Interference Suppressors – It's good to keep some ferrite cores and filters at your shack, to help you tackle interference issues rapidly and efficiently.
- Fuses – You never know when electricity will make up its mind to misbehave! Therefore, obtain spare fuses of all sizes and styles. At the same time, remember that you must never utilize a higher-value fuse for a smaller value one.
- Electrical Tape – You will need high-quality material (Scotch 33+, for example), if you wish to tackle outdoor connector sealing and other important jobs. If you have cheap electrical tape around, keep it handy for tackling throwaway or temporary jobs.
- Fasteners – Opt for a parts-cabinet assortment, comprised of #4 through #10 lock washers, nuts and screws. It is possible that some of your equipment is compatible with smaller metric-sized fasteners. Then, you will have to use 1/4-inch and 5/16-inch hardware, specifically for masts and antennas.

E. How to Maintain your Ham Radio Station

If you wish your ham radio venture to remain successful and trouble-free, you will have to perform regular maintenance. After all, you do the same for your vehicle, checkbook and so on! In this case, use a notebook for making references, notes and lists. Whenever you bring an additional piece of equipment into your shack, write it down. Other information that should go into it includes problems and their fixes, wiring of gadgets, details regarding inspections and checks, etc. Ensure that a date and time follows every entry. Over time, your notebook will suffice as a self-created encyclopedia for your ham radio station. It will also reveal the history of issues and solutions. Whenever you face similar problems in future, all that you will have to do is to open the notebook at the correct page and rediscover the solution that you had used in the past.

If you do not want to encounter too many headaches at any given time, it's best to set aside a specified time and day for carrying out investigations, testing and checking of individual components such as wires, masts, cables, power supplies, ropes, etc. You will have to be off the air when you do the maintenance. Therefore, choose carefully. You will have to do this on a regular basis, such that you are not suddenly entering a whirlpool of issues! The following checklist should make it easy for you to carry out routine maintenance.

Vacuum Clean the Shack

Yes, you heard right! You will have to drag your vacuum cleaner into your shack, whether you like it or not! It is imperative that you use it for cleaning your equipment and surroundings. This is because whenever dust, grime or crud uses your radio equipment as a residence, it prevents the heat generated by the equipment from dissipating. In fact, it takes the place of an insulator, restricting the flow of air in and out. As a result, the heat is stored inside the electronic components, unable to get out. Do not be surprised if various components stop functioning or refuse to function as efficiently as they should. The worst sufferers are circuits running on high voltage, such as a computer monitor, amplifier, etc.

Since you cannot remove the parts of every component and clean them from the inside, preventing dust in the first place, using a vacuum cleaner, makes sense. This way, your shack will be free of paper scraps, dust, bits of wire and all other kinds of junk. Otherwise, repairing equipment will turn out to be a costly affair. Remember to also clear your table of old magazines, loose magazines, coffee cups, etc. If you have empty cans of soft drinks lying around, recycle them. Finally, make sure that the ventilation holes and fans look spic and span and are free of blockage.

Check Everything Thoroughly

- **Ropes and guys** – Make it a habit to look at your surroundings whenever you walk by the places where you have positioned parts of your equipment. Consider the case of a tree branch rubbing against a rope; it can ultimately result in the rope breaking, causing your equipment to fall. Similarly, knots may loosen and open up. So keep an eye on things.
- **Antenna mounts, masts and towers** – Changing weather conditions can cause all manner of problems. The trick is to investigate everything with x-ray eyes before the weather turns bad in winter! Cart along a wrench with you, in order to test the tower. If you find any nuts or bolts even slightly loose, clamp them down. In case you find rust making an entry at some points, cover it up with cold galvanizing paint. Once the cold season passes by, make your rounds again.
- **Antennas and feed lines** – If it is possible to climb up to your antennas, do so. Otherwise, buy a powerful pair of binoculars. What do you look for? You must check for loose connections, if any. Also, note if

tape is unraveling, or something is twisted. Do you see any ties, damage to the cable jackets, etc.? If you do notice anything amiss, ensure that it is resolved immediately.

- **RF cables, switches, connectors and grounds** – Loose connections can result in all kinds of transmission and receiving issues. Therefore, check every single connector. If you find anything loose, tighten it. Remember that thermal cycling is at work here. Naturally, you cannot expect all connections to behave well all the time. It helps to cycle relays or rotate switches, in order to keep the contacts clean. Feedlines may fall prey to damage or kinks. Check them well. The ground connections must remain snug, always.

- **SWR on antennas** – Always be vigilant, specifically when you note changes in the frequency of even the minimum SWR. This may indicate the advent of connection problems, or water seepage into a particular antenna. If the SWR seems to move up and down all of a sudden, it indicates the presence of issues with the feed line or the tuning.

- **Transmitters and amplifiers** – You must check if the power output on all bands is full. In fact, use the full power output to check if the RF feedback on all the bands is coming through properly or not. Similarly, check the pickup on microphones, control signals or keying lines. Double-check the RF cables and antennas too.

- **Noise level** – If the feed lines are in great shape, the noise level will be normal. It is the same when the preamps are working appropriately. However, if you find that it is vacillating between high and low on all bands, then there is definitely something wrong. You may discover a new noise source, which is bound to cause you some amount of worry.

After you complete your maintenance check, make a list of all the things that need replacing or fixing. Use your station notebook to record everything. If you can figure out the exact reason for a specific problem, mention it in your notebook. This will help you to understand how to tackle the issue if it recurs in future. As you go through this exercise, you may even get ideas about some additions that might help your radio station to improve its performance. You may also wonder if you should make certain improvements in your way of operating. Whatever thoughts and ideas come to your mind, transfer them to your notebook.

As you continue with your regular maintenance checks, you will discover that certain things need regular updating. For instance, if your shack has a couple of moveable desks, you may have connections trailing along the ground. Considering that they are in a place that may be full of dust and dirt, you will have to pay them regular attention. Think of innovative ways to keep them safe, secure and dust-free. Regular checking will prevent issues like feedback, interference, etc., from cropping up. Therefore, do full-on monitoring trips three or four times a year. This way, you will not have to face unpleasant and sudden shocks.

F. Competent Troubleshooting

Do not assume that regular maintenance is the ultimate solution to all problems. You are dealing with electricity and electronics. There are bound to be failures and breakdowns sometime. Regardless, if you learn to recognize a problem rapidly, or even predict its advent in advance, you will be an efficient troubleshooter. Here are a few recommendations.

It is reasonable to panic as soon as you notice an issue with your radio, support equipment, etc. You might even jump to the conclusion that you will have to initiate big repairs and spend a lot of money. However, jumping to conclusions will get you nowhere. Remember that there is a solution somewhere, and you just have to find it! The first thing you should do is to take out your station notebook and see if there is a recording of a similar issue in the past. If not, and you are facing an entirely new issue, it does not matter. Get your thoughts down on paper first. This will help you to work through the problem in a systematic manner. It will also help you see the big picture and comprehend that some component is not functioning properly. Therefore, work downwards. From the big picture, work down, fragment by fragment, until you understand what you must do with your equipment.

If it helps, recheck everything from beginning to end once again. It does not help to make assumptions. You must obtain clarity about whatever you are facing. Then go through the manual that came with your equipment. There must be some advice in there, which the manufacturer felt would help you in times of crises. Using that advice, you may opt to make adjustments or changes to that particular gear, such that it begins functioning again. Write down a detailed description of what you did, in your notebook. Who knows, you may need it in the future! Additionally, if you discover that these adjustments or changes have not worked, you may refer to your notebook and reverse the whole process. The last resort is to consult a technician or professional and request repairs.

G. Troubleshoot Your Ham Radio Station

When you think about your station, or even look around, realize that it is a mass of equipment lying around, with every piece connected to the other. Now, if each item is to cooperate with you, it requires your cooperation in return. This means that its controls and connections will operate only when in alignment with certain conditions. Therefore, whenever there is a violation of these conditions, you should be able to trace the problem without using any test equipment. At the most, the humble voltmeter may come into play for checking the issue. Generally, the problem lies with either the equipment or its connection.

You may place your station's problems into two categories: RF and operational. RF problems include the absence of signals, high SWR or reports about the transmission/receipt of poor signal quality. Operational problems include the lack of communication between different pieces of equipment, inappropriate keying or lack of keying, or inability to turn on/off. Thus, whenever you notice any issue at your shack, see in which category you may place it. Of course, it is also possible that the problem does not fall into either category. Nonetheless, you will have to begin somewhere, if you have to find a solution.

RF Issues

RF refers to an electromagnetic wave frequency which lies between a range of 3 KHz and 300 GHz. This includes all the frequencies that people use for communications or sending radar signals. Now, sometimes, the RF does not reach the destination it is supposed to go to, and this leads to issues. You will be able to recognize the problem when you view a bad connector, cable or switching device. A switching device is a relay or switch. Sometimes, gear goes missing. Whatever the case, you will have to replace the equipment if you wish the radio to work properly again. Regardless, see if you can fix the issue, using the following recommendations.

- In case you have spares, use them to replace the adapters and cables.

- Thoroughly check the combinations of antennas and switching devices. Which combinations function well, and which have stopped working or are malfunctioning? It could be that the issue is common to a set of cables, a set of equipment or only a piece of equipment.
- The antennas' feed lines deserve some attention too. Does your antenna feed point have a DC connection across it? This connection could be an impedance matching transformer or a tuning network. The problem may relate to this connection. Yagi beams and quad loops, which are beta matched, reveal a feed point that has less ohms of resistance traveling across it. In contrast, Yagi beams, which are gamma matched, display an open circuit. As you are pondering this problem, do not forget to note down the normal values of these resistances in your station notebook. Whenever you troubleshoot, the comparison will prove helpful.
- Maybe you can bypass relays, filters or switches, to see if this will fix your problem. However, this is a temporary solution. Alternatively, you can jumper around the relays, filters or switches, albeit temporarily. In case the jumper is not visible, make a note, such that you remember this point.
- If the problem is with the test gear, use the noise from the receiver to resolve it. For instance, you can disconnect all the cables from the antennas and reconnect them once again. This will bring about visible changes in the noise level.

You may encounter other problems too, especially concerning microphones, enclosures for equipment or interference. To illustrate, your microphones may be storing too much heat within them. You will understand when the metal microphone case causes an RF burn on your lips! With regard to your equipment, their enclosures may not be healthy for their functioning. Or your computers and accessories may be experiencing too much of interference. Maybe these issues just need a little bit of grounding. If so, try these actions:

- Are you sure that all your gear is grounded to your radio station's RF ground bus? It does no harm to check everything once again, even if you are sure that everything is all right. If your equipment is indeed secure, you can be at peace!
- In case the grounding is not appropriate, you might opt for coiling up an extremely long cable or devise diverse grounding connections.

- If you own gear that is difficult to ground, you might add beads to the cables connecting to it. An alternative to beads is ferrite RF suppression cores.
- The cables connected to audio equipment or controlling gear tend to be rather delicate. If you have this habit of yanking upon them, or flexing them, you risk breaking them. Therefore, check the shield connections on these cables, to see if the cables are safe or not.

If you are a fan of the higher HF bands, such as 21, 24 and 28 MHz, you should be able to observe that your connections are akin to antennas! This is because their lengths are beyond 1/8 of a wavelength. To illustrate, if you have a six-foot serial cable in place, you will notice that it is around 3/16 wavelengths in length. Since it functions on a band of 28 MHz, even the RF voltage is quite sizeable at its midpoint. This is astonishing indeed, since you have taken care to ground both ends! Therefore, if you find that a lone band is experiencing RF pickup problems, see if fixing a counterpoise wire, measuring 1/4 wavelength, will resolve the issue. The idea is to divert the RF hot spot from the concerned equipment.

You may not be aware of it, but a quarter-wave conductor, which has an open circuit at one end, may appear as if it has suffered a short circuit at the other. Therefore, when you fix the counterpoise wire to the enclosure that houses the malfunctioning equipment, the RF voltage may come down. In turn, the interference reduces or suffers elimination. It is imperative to keep this counterpoise wire well insulated. Make sure that its open end does not lie near other equipment. Above all, ensure that nobody comes near it.

Operational Issues

We may place them in three categories, such as control, power and data. Therefore, prior to actually tackling any issue, it is important to see which category it falls into.

Issues with Control

Surprisingly, you may be the mastermind behind your control problems! If yes, it should prove very difficult to fix issues in the absence of expert help. Of course, you may opt for trial and error methods, but you have to be sure that you do not actually compound the existing issues, instead of resolving them! Alternatively, the problems might be the outcome of errors in control input. Now, how do you find solutions to control problems, regardless of the culprits that caused them? The following steps might help:

Step 1

To begin with, recheck your operating controls. Are you sure they are all set properly? After all, it is quite easy to mistakenly move a control, or cause it to move by bumping into it. Check the operator's manual, to find out if the setting for each mode is correct or not. Then, opt for systematically setting up the various controls once again. Remember that there are some controls under an access panel, as well as on the back panel too.

When you are going over your radio, pay close attention to the controls on the front panel. For instance, there is the Squelch, which is capable of muting the audio suddenly and unexpectedly. Then there is MOX, which is responsible for switching on the transmitter. The receive antenna will help the receiver to go dead in the absence of a receive antenna. Once you have observed all these controls closely, you should be able to figure out the problem. However, there is a last resort too. You may go for pressing the button for a hard reset. This will restore all the factory default settings. The snag is that your radio will lose all its memory settings in the bargain.

Step 2

It is time to disconnect every single cable from your radio, albeit one at a time. Just allow the power cable and the antenna to stay on. Which is the cable that is responsible for the signals causing the problem? This is your priority, before you move on to other cables. Observe the behavior of each cable. If you observe a behavioral change with any cable, go back to the operator's manual. Study all the information related to that particular cable. Is it possible that the signals passing through that specific cable are responsible for the issue? Then, use an ohmmeter for checking intermittent connections or shorts, which may be present in that cable. All that you have to do is to wiggle the connector, while observing the meter.

Step 3

It is possible that your equipment is refusing to respond to PTT or keying, which are the control inputs. If this happens, you will have to simulate the control signal with the aid of a switch. Control signals are majorly contact or switch closures. These closures occur between the ground or 12V and a connector pin.

At the outset, bring out a spare and replace the control cable with it. The connector must be unwired. Next, bring out a test lead (this refers to wires, which have small clips at each end), such that you may jumper the pin to the voltage. In case the pins are located close to one another, you might consider soldering a small switch to the connector, with the aid of short wires. Try making the connection manually, in order to see if the gear gives the proper response or not. If it does, it indicates a fault in the device or cable responsible for generating the signal. Otherwise, the fault lies with the equipment that you are testing. Nevertheless, by now, your knowledge of electronics should help you to link the problem with a particular piece of equipment. Now decide if you can conduct the repairs by yourself (with the aid of the operating and schematic manual), or not. If it all seems too worrying, maybe you should have someone more experienced step in.

Issues with Power

No power is a highly obvious fault, since your equipment will not work! If the problem is with the failure of the high voltage supply, that's a big deal! A more subtle problem is poor connections, an AC ripple or a slightly high/low voltage. Whatever the case, remember that power is powerful, and you should never take it lightly! To illustrate, even if you see the power supply light staring steadily at you, don't take it for granted that the output is functioning at the right voltage. Therefore, monitor the power supply voltage first, whenever you sense a problem. Here are some fixes to your power problems, which just might work beautifully!

If you see your radio behaving strangely, you might give a thought to the output. It could be that the output is low in voltage. As a result, the microprocessor will not function appropriately. Do not be surprised when you view incorrect responses to controls, bizarre displays or the loss of external control. It is also possible for low voltage to reduce the stability of RF by producing raspy, chirpy or drifting signals. A third possibility is low power output.

Then again, it could be that the equipment is responsible for the problem. The power supply may not be at fault at all. However, you must be sure that this is the case. Unless you are sure, do not change power suppliers. Do not try to connect the power supply to an input or a shorted cable. This will rapidly destroy the supplier's output circuits. Similarly, if a circuit breaker or fuse continually opens up, do not try to jumper it. Try to discover why this is happening, first.

Do you feel that the connection is poor? If so, measure the voltage at the load first. Here, the load might be the radio. The voltage does drop under load if poor connections exist in a connector or cable. It becomes difficult to isolate these connections, since they become problematic with high currents only. You may see such high currents while transmitting. In contrast, the voltage may be excellent while receiving signals. Therefore, when you see the indicator light going dim most of the time, you may be certain that the supply of power is failing or that poor connections are present.

Try checking the power supply with both an AC and DC meter range. If you hear a hum on your signal, it means that the power supply is failing. In case your equipment is battery-powered, the hum indicates the failure of the battery's power. You may use a DC voltmeter to see if the power supply output shows less than 100 mV of AC, or not.

Please note that it is dangerous to work on a power line functioning on AC. It is equally dangerous to work on 50-volt or higher supplies. Therefore, always consider safety first, prior to trying anything foolhardy.

Issues with Data

This is becoming a common problem at modern ham radio stations. In general, RS-232 connections come into play, while creating interfaces between radios, computers and data controllers. However, even after installing new equipment, if you are unable to make it compatible with your other equipment, then you will have no choice but to blame any of the following culprits.

Baud Rate

This is responsible for specifying how fast you may send your data. It is the data framing parameters of parity, start bits and stop bits. These framing parameters are responsible for specifying what kind of format should link with each byte of data. It is possible to set them in alignment with a software interface or via switches that exist in an accessory. However, if the baud rate is not functioning properly, it will ensure that even the links do not operate properly. It does not matter that the wiring is correct. This will still happen.

Errors in Protocol

They show up as mismatches in the version or type of equipment. To illustrate, if you buy a personal computer possessing a Kenwood radio control, it won't be able to take charge of a Ten-Tec radio or a Yaesu radio. Therefore, it is imperative for you to ensure that all your equipment is compatible with the same protocol. Alternatively, the equipment must be able to work with the exact models that you have with you.

Inappropriate Configuration in Wiring

Prior to reconnecting the requisite control signals, it's best to read the operating manuals for both types of equipment. It could be that you will need to provide jumpering pins to the cables. Check whether the pins must go only at one end, or at both ends. The null-modem (common cable configuration) helps in bringing together all RS-232 output signals and their companion input. It is your choice whether to make null-modem cables on your own or purchase them. Alternatively, you may opt for a straight-through cable (pin 1 linked to pin1, pin 2 linked to pin 2 and so on). Fix a null-modem adapter at one end of the cable.

You may view RS-232 as three configurations of software. These configurations are quite common. The first one is three-wire configurations, which take charge of controlling the flow of data. The three configurations are Receive Data or RxD, Transmit Data or TxD, and ground.

The second type is five-wire configurations, wherein the control lines take up the responsibility of implementing hardware handshaking. These five configurations include RxD, TxD, ground, RTS (Request to Send) and CTS (Clear to Send). The handshaking is to ensure that there is some control over the data flow. If the control lines don't have appropriate configurations, the output port will not be able to relay data at all.

The last type is seven-wire configurations. The two additions to the existing five include DTR (Data Terminal Ready) and DSR (Data Set Ready). If you have to confront CD, or Carrier Detect, too, it means that you are dealing with packet data systems. This would suggest that the channel is busy.

Then again, it could be that you discover that a piece of equipment, which was engaged in good communication before, has suddenly failed. When you check for the cause of the problem, you might find that the configuration of the software has changed at one end of the link. Alternatively, a cable may be loose too. Whatever the case, do not panic. Just double-check the swap cables, as well as the communications settings.

Fortunately, you have several options for wiring the RS-232 link. You might even find some tutorials connected to RS-232 interfaces. There should be some handy websites online, which you should find easy to follow, since they display diagrams.

A wonderful tool, the breakout box, will help you troubleshoot all kinds of serial data connections. With the help of this box, you may monitor the existing status and ongoing status of all the control lines and data. If necessary, you can disconnect the jumper lines, as well as the disconnect lines. It becomes easy to determine if there has been swapping of the transmission and reception lines, or not. You should also be able to find out if there has been misrouting with a control signal, or not. The RS-232 status testers are not as capable as the breakout box is, but are definitely handy in times of need. These testers reveal the status of every commonly-used line in place.

Computers from 2004 onward only have FireWire serial data or USB ports. However, RS-232 remains the serial data interface standard in every ham radio shack. Thus, you have two options. You may have a multiport RS-232 interface board going into your computer. Alternatively, you may opt for USB-to-RS-232 adapters. Whichever you opt to use, you must first, install the requisite drivers in your computer's operating system. These interface boards exhibit maximum compatibility with radio gear. USB adapters take up the job of converting the data not once, but twice. The first time, it is between the USB and RS-232. The second time is between the USB and the host computer. As a result, there are greater chances of incompatibility. Therefore, it would best to consult the vendor selling your accessories about what adapters he/she would recommend.

H. Dealing with Interference

There is no denying that whatever problems you might face outside your shack, the primary culprit is often RF interference. You will know when your RFI conveys messages about someone being able to hear you on their telephone, or complains about their garage door moving up and down without stopping! Obviously, everything sounds bizarre. Equally obviously, your ham radio station is confronting interference from another electronic or electric device, albeit from an unknown location. It is not easy to resolve these issues, but it is not impossible either!

You might begin by locating the information page on RFI at the official ARRL website. Alternatively, you may read *The ARRL RFI* book. It depicts every source of interference. Of course, this does not mean that other official websites are useless. It is just that everyone continues to see ARRL as the best or standard in the world of ham radio operations! You may also ask members at your club for expert assistance. After all, there is no denying that all ham radio operators face the challenge of occasional interference at some time or other during their ham life! Just tackle the issue as well as you can, using all the resources at your disposal.

Handling Interference to your Equipment

Two kinds of interference are likely to take place outside your shack. One is electric, and the other is electronic. The electric interference may take the form of occurrence in arcing in equipment or power lines. The equipment that comes into play here includes electric fences, motors and heaters. The electronic interference is the outcome of RF signals leaking from computers and consumer appliances that are in operation in the nearby vicinity or transmitters operating close to your ham radio station. You will be able to recognize either one of them via their distinctive signature or characteristic sounds.

Identifying Sources of Electric Interference

Power Lines

You should hear an intermittent or steady buzzing at a frequency of 60 Hz or 120 Hz. It could even be that the weather that is having an impact on interference.

Regardless, what could be causing this kind of noise emission? It could be due to a corona discharge or arcing in the power lines. An electric field exists between the surfaces of two conductors. This electric field may go beyond a critical value. This self-sustaining ionizing discharge is visible between the surface areas lying closest to one another. We call the surface areas *corona* and the discharge *corona discharge*. The corona discharge causes electric arcing. In turn, the arcing may cause an electric fire.

However, when does arcing or corona discharge occur?

If the insulators are cracked or dirty on the inside or on the surface, they may become responsible for the occurrence of arcing. Sometimes, the surfaces of two wires come into friction with one another. One may be the ground wire, while the other is the neutral wire. With regard to corona discharge, it occurs if the air molecules undergo ionization, leading to electrical leakage into the atmosphere. The discharge is visible at the high-voltage points that are present on sharp objects. Because of corona discharge or arcing, you should be able to note the interference as a buzzing noise. This is because the occurrences depend on the peaking of the 60 Hz waveform. This is the 120 Hz waveform.

Whatever the case, do not try to fix issues with power lines on your own, unless you want to be electrocuted! Just ask the power company to take over. When they arrive, you may provide assistance in locating the equipment at fault. It is possible to track the source of noise with the help of an AM radio that uses a battery for power. It is also possible to bring a UHF/VHF handheld radio, possessing an AM mode, into play as an alternative. The aircraft band works well on the handheld radio. Then again, do you have a rotatable antenna at your shack/home? It should help in pinpointing the direction of the noise. Note that the null off the side of your beam antenna is sharper in comparison to the peak of the pattern.

Walk/drive alongside the power lines, in the direction of the noise, just to discover the exact region where the noise is at its peak. Driving around in a car with the AM radio tuned between stations is a great strategy for discovering power poles whose hardware is not functioning properly. Whenever you suspect that a specific pole is malfunctioning, note down the identifying numbers displayed on it. Several companies may have been involved in setting up a particular pole. Note down every single contact detail, and get to work contacting them. We reiterate, with regard to interference to your equipment from these power lines, consult your utility and request that they report it to the concerned authorities.

Electric Fences

Are you hearing certain noises at quick one-second intervals? Do they sound like pop pop pop? If the answer is in the affirmative, you may wonder if there is a defective charger somewhere around the place. However, the culprit is generally arcing occurring between the wires and weeds, ground or brush, at the electric fence. The fence behaves like an antenna, radiating the radio noise, which the arc generates. An alternative reason for interference could be the presence of missing or broken insulators, due to the advent of adverse weather conditions. This kind of interference affects television and radio reception for quite some distance in the nearby vicinity, sometimes stretching to as far as a mile. If you have AM radio, you should be able to hear that distinct tick-tick-tick sound.

At the outset, locate the source of the fault. Is there a spark gap somewhere? Sometimes, the spark gap occurs at the bad splices in the fencing or the gate hooks. Then again, there could be too much vegetation around the place because you have permitted the weeds to rise as high as the height of the electric fence. Therefore, whenever the branches connect with the electric fence, a short occurs, resulting in significant noise. Soon enough, the plant burns, thereby breaking the connection. If you continue to allow the weeds to grow back, the same process recurs, causing interference.

When you find out what is causing the interference, you may take necessary steps to rectify the fault. It is possible to use a systematic method to stop this irritating noise by referring to genuine and knowledgeable articles and handbooks. However, it is a lengthy process. Therefore, if you lack the patience to go through the entire process, just call in an electrician or a professional to conduct the necessary repairs.

Industrial Equipment

At first, you may mistake the interference for noise resulting from faulty power lines. However, the pattern here seems to be more regular in nature. This is what you observe with faulty heaters or motors, which operate in accordance with a cycle. To illustrate, think of the furnace fans, vacuum cleaners and sewing machines which you use at home. When you observe this pattern, you may be sure that there is something wrong with the industrial equipment.

Dimmers and Speed Controls

Whenever you bring motors or lights into play, there is a low-level noise which repeatedly appears and disappears. It is akin to what you hear from power lines.

Defective Contacts

If your ears feel irritated by rasping noises or highly erratic buzzing sounds, you may be sure that switches/thermostats, dealing with heavy loads, are heading for failure. If you do not recognize and fix these defective contacts immediately, you are risking a fire at your shack or home.

Automotive Ignition Noise

If there is an arcing in the ignition system, it is going to result in a kind of buzzing that varies in alignment with the varying in engine speed.

The first thing to do, whenever you find something amiss with equipment functioning on electric power, is to find out whether the faulty device is present on your premises or someone else's. You should find it easy to track the source of noise. Keep note of the recurring pattern, whenever there is some noise. Then, connect this noise to a particular appliance. An alternative method is to turn off the circuit breakers at home, one by one. This will help you track the circuit which is providing power to that particular appliance/device. Check all the devices/appliances connected to that particular circuit.

Sometimes you may feel that the noise is coming from outside your home or ham radio shack. Therefore, try to figure out the direction from which the noise is originating. Walk/drive that direction with a portable receiver in hand. You might request professional help from websites or authentic articles on how to approach the owner on whose property the interfering device is located.

Identifying Sources of Electronic Noise

Even electronic devices emit their signature sounds when malfunctioning.

Leakage from Cable Television

You should hear that odd buzzing, which is a video signal, or a FM signal, which is an audio signal, emanating at UHF and VHF. This happens even when there is a program on display. Then again, your television set may suffer interference on that specific cable channel. This indicates a definite electronic leakage.

Cable and Power-Line Modems

Tune in to the HF bands, and be astonished by the hisses and rasps emanating from them! Alternate sounds are warbling or steady tones.

Entertainment Systems, Appliances and Games under the Control of Microprocessors or Computers

A single frequency should suffice to suggest interference because of the warbling/steady sounds emanating on the band. The interference is strongest on HF. However, it is possible to discover it on UHF and VHF too.

Broadcast Interference

In this case, there may be an overload on your receiver. There could also be the combining of strong signals, resulting in the generation of spurious signals. Sometimes, the broadcast transmitter emits a spurious output. As a result, you should be able to obtain a steady AM signal along with the programming. Alternatively, you may hear bursts of speech. Then again, there could be data interference.

With regard to every kind of electronic interference, you will have to practice a different technique for discovering the source. In case the interference is from an electronic source at your home/shack or close by, the signals will be weak. In case the interference is originating from a source outside your property, you will have to use a portable receiver in order to listen to it carefully. After discovering the source, you will have to approach the owner of the property in a tactful manner. Nonetheless, it is not too difficult to eliminate interference emanating from electronic sources. Apart from ARRL, there are several professional organizations which offer detailed information and guidance, as well as the services of technical experts. Use this knowledge to conduct your own investigations, as well as effecting all the necessary repairs by yourself. Interference suppression techniques are easy to understand, provided you retain a cool mindset and decide to work through each problem without giving up!

Ferrite is Useful

As mentioned earlier, ferrite, a magnetic ceramic material, is a great suppressor of interference. You see this core in transformers and RF conductors. The material may take the form of toroid cores, which are rectangular and circular rings. It may also take the form of rods or beads. Beads refer to small toroids, which are able to slip over wires. Ferrite possesses healthy magnetic characteristics at RF. It is available in diverse formulations or mixes that mixes permit it to work at optimum level for diverse frequency ranges. To illustrate, if you wish to use ferrite for HF, you must opt for amidon ferrite, which is comprised of Type 73 material. Similarly, if you want something for VHF, opt for the Type 43 ferrites.

Preventing RFI or radio frequency interference means that you want to reduce the flow of unwanted RF current. Inductors possess an enhanced resistance to the flow of alternate current (AC). In technical terms, this is enhanced reactance. Therefore, if you bring an inductor into play at the designated spot, you will have found a good solution.

How do you make the inductor?

You may opt for twisting a cable or wire into a coil. However, you will need sufficient inductance. Therefore, you will also need a sizeable coil, measuring below 10 MHz. The solution to this lies in using a ferrite core. Just wind the wire on this core. Alternatively, you may utilize a ferrite bead. You will then be able to produce sufficient inductance, even in an extremely small volume.

Since ferrite beads and cores are rather small, it is easy to position them extremely close to the point where the unwanted signal is making an entry or an exit into a particular piece of equipment. Beads may slip over cables and wires easily, because of their compact design. Therefore, if you want to have more inductance, use more ferrite beads. It is easy to secure them to the specific cable via tape, heat-shrink tubing or plastic cable-tie. As for ferrite cores, wind them over the concerned cable or wire, using several turns. This is an excellent technique for dealing with power/telephone cords. Then again, you may have access to split ferrite cores, which possess a plastic cover for keeping the cores together. As a result, you may place the split cores on the concerned cable and begin winding them around it. However, you must make sure that the cable already has a large connector in place, before you begin your winding job. This will make your task easier.

Handling Interference to Other Equipment

It is important to make your own premises free of any kind of interference, before you offer your services elsewhere. If you are a low-power UHF/VHF operator, you need have no worries. Otherwise, you obviously own at least a lone appliance, which reveals diverse reactions to your transmissions. These reactions could take the form of clicking, buzzing, duck imitations of your speech or humming. In fact, you might feel that the appliance is behaving like an extremely unselective AM receiver. You also wonder if your strong signals are converting to audio, similar to what used to happen in the old crystal radio sets. Nevertheless, do not blame the ham radio for what is happening. The appliance is unable to reject your signal, thereby proving to be rather annoying in behavior.

What should you do?

Well, your major goal should be to keep the signal away from your appliance, such that it receives no signals at all. Towards this end, begin disconnecting all the wires and accessory cords. Check if the problem has disappeared. If it has, return the wires and accessory cords to their respective places, one at a time. You must know which is behaving as the antenna. For instance, speaker leads and power cords are excellent antennas, assisting in conduction of RF into the concerned appliance. You may try something else too. Take some candidate cables and wind them onto a ferrite core. Place the core close to the concerned appliance. You may have to link the core to all the leads, despite just one or two proving sensitive. Check if this is the actual solution to the problem. In case the device/appliance functions on a battery, and exhibits no leads, you may have to conclude that you cannot fix the problem. You might even have to consider replacing the device/appliance, or simply get used to the interference. See if the manufacturer's website or manuals have any solutions. Alternatively, consult members of your club, official websites of other ham radio operators, etc. Finally, key in the model number of your device/appliance and the word *interference* into a search engine like Google, and wait for the results.

There are certain devices/appliances which are favorites of interference.

Alarm Systems

You feel happy seeing the lengths of wire strung all around your premises. You feel even happier seeing the connections between these wires and the diverse switches and sensors. This is because the whole picture seems to grant the impression of a splendidly functioning antenna! However, there is an enemy lying in wait to ruin your happiness, in the form of the system controller! It tends to be confused by the RF, mistaking it for a sensor trip. Therefore, it would be best to opt for system installers, who bring along interference suppression kits and tackle several issues very competently.

Touch Lamps

They are inclined to be friendly with any strong signal that comes their way, on any frequency! Even if you opt for placing ferrite cords on the power cord, you cannot hope to gain completely satisfactory results. In fact, the results are generally mixed. If you wish to, you may try internal modifications. However, the simplest and best thing to do would be to replace them with new ones.

TVs and VCRs

The UHF/VHF cable input is the culprit that permits undesirable RF to enter the TV or VCR. As a result, the highly sensitive tuners experience an overload. In case the interference occurs from an HF signal, you may utilize a high-pass filter and resolve the issue. Then again, the cable installations may display loose connections, permitting ham signals to enter the set. For instance, when you are transmitting on the two-meter band, interference may occur with cable channel 18. Place a call to the concerned cable company and request that they tighten all the connections.

Cordless Telephones

Cordless phones that are inclined towards 47 MHz frequencies prove to be excessively sensitive to powerful out-of-band signals. Recently, models which use 900 MHz and 2.4 GHz radio links are replacing the 47 MHZ. They are not as sensitive to RF as their earlier counterparts were. Therefore, do not think twice when you come across a 47 MHz unit. Simply obtain a new cordless telephone in its place!

Do not hesitate to use your electronic equipment at home as your 'guinea pigs'! As you gain expertise in diagnosing and fixing interference problems, you may offer your services to others in the neighborhood too.

I. Rules for Part-15 Appliances

The Federal Communications Commission (FCC) has come up with Part-15 Rules for unlicensed devices. These devices include garage door openers, cordless telephones, wireless headsets, wireless modems, etc. The rules are applicable only if they are using RF signals to communicate or operate. Even though computers and video games radiate RF signals unintentionally, they are subject to the same rules. As per the tradeoff generated by these rules, the owners of such devices need not purchase licenses. However, they should take care to see that they do not interfere with a service that owns a proper operating license. One such service is the amateur radio. Similarly, these devices may not accept interference from licensed services either. Thus far, the agreement has worked quite well. If there are flaws, they have occurred near a ham radio shack that boasts of strong transmission or highly sensitive reception. Obviously, the FCC has to initiate further discussions in this matter.

J. Build/Not to Build

There is something wonderfully exhilarating about creating something with your own hands and tools! Therefore, you could try building some pieces of your equipment too, such as a speaker switch, an antenna, etc. There is plenty of guidance available both on and offline. In fact, it is something that many amateur hams seem to follow as a tradition!

Use a Kit

When you begin your construction activities, you gain experience in operating various pieces of equipment, repairing them as necessary and maintaining their good health. However, since you are new to the concept of building, you might use a building kit. They come in all sizes and shapes, and you may get them both online and offline. True, you may feel overwhelmed by all the brands and specifications on display. However, it would be best to opt for an extremely simple kit. That will help you practice until you gain more confidence, before you reach for higher-level packages.

When you opt for a kit to build a few simple devices, you actually save on money. You also gain the opportunity to add several test instruments to your workbench. It does not matter what materials you utilize; the results should be splendid indeed! Who knows, you may grow so much in confidence that you decide to build your own radio someday. Towards this end, there are good-quality radio kits available to help you achieve your ambition. Just make sure you have a great set of soldering tools and skills handy. Always, handle your kit as a master would, for you should be proud of your resulting achievements. Keep the building and operating manuals handy. You must also study the schematics when you buy a high-quality radio kit. It is imperative that you understand the mechanics behind building a radio. Only then will your controls and front panel look professional!

Do you feel confident enough to begin a construction project from scratch? All that you have as assistants are some instructions from an article and a blank notebook to jot your ideas down. However, you have no kit that you can depend upon here. You will have to make your own! Keeping this in mind, opt for a simple project in the beginning. Just copy the diagram of a circuit outlined in a handbook or ham magazine. Obviously, the diagram will have an accompanying article about how to assemble this circuit, test it and make it work. After obtaining a blank and printed circuit board, begin construction, keeping the instructions, schematics and list of components in front of you. It will help if you convince your imagination that you are preparing the circuit for someone else! In case there is an assembly drawing on display, photocopy that. In fact, it would help to have several copies of the circuit, such that you can mark every step, soon after completing it. Ensure that you do not miss any critical steps.

When you obtain your components, sort them out by value and type first. Then place each component in an old muffin pan or a jar. Keep a notebook handy. Any notes you take should prove quite useful in future. After all, every aspect of your building, modifying and testing should go into the book as a new entry. These entries may include things like labeled schematics, calculations, original values, photographs, tests undertaken and so on. Thus, at the end of it all, you should be able to see a completed circuit in all its glory!

Chapter 4—How Do Radio Signals Propagate

The equipment that has the greatest impact on the way the radio station performs is the antenna. Therefore, it's no surprise that this is a ham operator's favorite subject. For this reason, there is a huge variety of antennas available for all types of radio services. However, to select the best antenna for your specific purpose, it is important to understand the underlying concept of propagation, which means traveling from one place to another.

How Do Radio Signals Travel: Propagation

When radio waves are transmitted from the antenna, they move straight unless they are distracted or reflected by an object along the way. This behavior can be compared to that of light waves, which are an extremely high frequency form of radio waves. And, just like light, when radio waves travel long distances, their strength decreases as they move away from the transmitting antenna. There comes a point when the signal strength becomes so weak that it cannot be even received by another circuit. Therefore, care must be taken that the signal is received before it dies off completely, and this distance over which it can be received is known as the range.

When transmitting and receiving antenna are in line with each other, the radio signals travel straight and the process if known as line-of-sight propagation. Most of the propagations that take place in the higher frequencies are known to be line-of-sight. This doesn't mean that radio waves at lower frequencies cannot travel straight; they can travel along the earth's surface, in a process known as ground wave propagation. The distraction in the path of radio waves can include any sudden change in the medium through which they are traveling, such as hills, mountains, buildings and so on. When radio signals get these distractions, they create shadows, especially in the case of higher frequency signals such as VHF. The edges of the objects that come in the pathway of these signals diffract the radio waves, which leads to knife-edge propagation. Another type of propagation is refraction, which is caused due to bending of radio signals in the medium they travel. The process of refraction is used to receive signals at specific points on the earth. When the radio signals start bending, they are extended in a backward direction, towards the ground, so that this refraction counteracts the earth's curvature and is received at points that are beyond the line-of-sight horizon.

Radio waves are also capable of penetrating solid objects; however, the opening in these solids should be more than half a wavelength. Due to this math, the shorter wavelength signals, traveling at lower frequencies, are more effective into and out of the buildings.

Another common phenomenon that takes places around radio signals is multipath. Radio signals get diffracted, refracted, etc. taking different paths in the atmosphere after being transmitted. When they arrive at the receiver, they interfere with each other. They are in and out of phase. When they are out of phase, they cancel each other, leading to a phenomenon known as multipath. When this happens, the signal becomes too weak and distorted. One way of preventing this is moving the antenna a few feet so that the interference of radio waves doesn't happen in a way that the resulting signals cancel each other at the receiver. Even when the radio signals are transmitted from different stations, this phenomenon can lead to weak reception although the receiver is good enough. Due to the same reason, the mobile signals traveling from a mobile station undergo fading in an area where multipath is present, and the process is known as mobile flutter. When digital signals undergo multipath, the received signals will have a higher error rate.

The propagation of signals above and at VHF frequencies that are assisted by phenomena happening in atmosphere (temperature variations, weather, etc.) is known as tropospheric propagation. Even the different layers of atmosphere, that bring along with them their own unique characteristics, guide these signals to go a long way. This kind of propagation, caused due to unique atmospheric characteristics, is normally used by hams to create contacts at these higher frequencies, which are otherwise not possible in the case of line-of-sight propagation. Even the conductive elements that are present in the atmosphere are capable of reflecting radio signals, the details of which depend on the nature and size of the reflecting surface. This is because each frequency acts in a different way when it comes in contact with the reflectors; for instance, aircrafts can reflect the signals for hundred miles.

Air in the lower layers of atmosphere is relatively dense. The upper layers, where there is not much air, are known as the ionosphere. This is spread across from 30 to 260 miles above the earth surface, where nitrogen and oxygen are exposed to highly energetic and intense UV rays radiated from the sun. The rays are so intense that they create positively-charged ions from the gas by rejecting the negatively-charged electrons. As this leads to the creation of ions and electrons, a conductive region is created high above the Earth. These ionosphere layers refract and absorb the radio waves, depending on the time of the day.

Another propagation type, known as sky wave propagation, or skip, is when radio waves at higher frequencies bend completely towards the earth after being refracted in the ionosphere layers, as if they have been reflected. Being conductive, even the earth's surface reflects the radio waves which makes them travel back and forth between the ground and ionosphere several times; this ensures that they travel through long distances. This is the common method that hams use to make long-distance contacts. Each time a signal gets reflected from the ionosphere, it is known as a hop.

How the ionosphere treats a signal also depends on the radio wave frequencies. Signals at higher frequencies bend more as compared to those at lower frequencies. Therefore, all those signals that are at VHF and higher frequencies bend just a little bit and get lost in space; that's the reason these are hardly ever heard if repeaters are not used. The highest frequency made to reach back to the earth surface through reflection is known as the maximum usable frequency (MUF) between the transmitter and receiver. The lowest possible frequency that can be sustained between these two systems, without fading away, is known as the lowest usable frequency (LUF).

The UV radiation becomes more intense as the sunspot activities increase, which, in turn, also increases the maximum level of ionization that takes place in the atmosphere. This increases the MUF between the two stations, and, during the days when we have maximum solar activity from sun, the higher frequency bands open up right after sunset. The band is said to be open when sky wave propagation is possible between the two points. If it is not, the band is closed. And since the ionization depends on solar radiation, areas have different ionization on top of them during the daylight as compared to during the night. This means radio propagation takes place in certain directions, while in other areas it doesn't, depending on which bands are open and which are closed due to the earth's movement.

Experts who study the impact of various elements on the signal at VHF and UHF find the ionospheric propagation to be very interesting. When sun's radiation is intense, the layers of ionosphere can bend even the VHF signals up to an extent that they reach back to the earth. And when these bands are open, they help in making long-distance connections that are not possible under normal circumstances. Also, certain parts of ionosphere become ionized enough to reflect the higher-frequency signals back to earth, commonly known as sporadic propagation. When this propagation takes place, ionosphere reflects various radio signals.

One amazing example to explain this phenomenon is that of the aurora—the northern lights that are the result of thin layers of charged particles moving through the lower layers of the ionosphere. And since the charged particles are constantly changing, the strength of the radio signals varies with them. This region is also home to various types of meteors. We often read and hear about these meteors, which are the result of meteoroids burning up in the upper layers of atmosphere. They are known to have tails, which are actually ionized gas, that last for a few seconds. Certain signals get bounced off from these meteor tails, and the process is known as meteor scatter propagation.

Antennas Play a Key Role in Propagation

Antennas play a key role in signal propagation. Most people feel that an antenna is a specific object, but the fact is that an antenna can be any electric conductor that can help in propagating radio signals; it can be a pipe, any wire or even a part of your car. The important thing is, for an antenna to transmit and receive the radio signals effectively, its dimension should be such that it is an appreciable fraction of the wavelength of signal that has to be propagated. And if you are wondering how the radio signals reach an antenna, it is the job of a feed line. A feed line delivers the signals to the antenna and, from there, it is propagated. The point at which the feed line is connected to the antenna is known as the feed point of that antenna. And if we go by the basics that we discussed, impedance being the radio of voltage to current, the feed-point impedance of the antenna is the ratio of radio signal voltage to current at the feed point. The antenna is resonant when this impedance is only resistance, without any reactance. This antenna impedance depends on its physical dimensions and the wavelength of the signal at a given frequency. And, as frequency varies, the impedance also changes because the wavelength of the signal changes; however, the dimensions of the antenna remain the same. Other factors that impact the impedance of an antenna's feed point include the conductors available in the surroundings of the antenna and how high it is placed from the ground level.

Elements: An antenna by itself doesn't transmit or receive the signals; it's the conducting parts of the antenna that do this job—these parts are known as the elements. When an antenna is made up of multiple elements, it is known as an array. The element connected to the feed line is known as the driven element, as it is the one that helps pick up the signal from the source. If there are multiple elements connected to the feed line, it is termed a driven array. There are also elements that are not in direct touch with the feed line but still play a role in governing the performance of the antenna. These elements are known as the parasitic elements.

Electromagnetism: Antenna elements carry the radio frequency current, resulting in propagating the radio signal away from the antenna. This wave is comprised of both electrical and magnetic energy created by the electrons that move up and down in the antenna. They can be visualized like the ripples traveling back and forth in the water. Since both electrical and magnetic fields are present in the radio waves, these are known as electromagnetic waves. These fields oscillate at the frequency of the current present in the antenna. When any radio signal passes by an antenna, the electrons present in the elements experience both the electric and magnetic force in it. These are caused due to the electric and magnetic field of the radio wave. The electron then oscillates at the frequency of the radio wave signal. It is because of this frequency that it oscillates at, and these forces that it moves in the element, back and forth, which result in the radio frequency current in the element.

Polarization: Another important phenomenon that plays an important role in the propagation of RF waves is polarization, which refers to the orientation of electric field in a radio wave. If an antenna is horizontally polarized, a radio wave whose electric field is horizontally oriented is reflected, and the antennas that are vertically polarized have an electric field in a direction that is perpendicular to the earth's surface. Maximum transmission of the signal by the antenna happens when the electric field of the wave and element of antenna have the same polarization. This is because it is the electric field that causes the electrons to move back and forth in the element of the antenna, resulting in the creation of the current and the signal in the antenna. This is the reason why people often hold radio signals in a way that the antenna is in line with the antenna of the receiving station. On the other hand, when the polarization of a receiving and transmitting antenna is not aligned properly, the signal is not properly received, resulting in a weak signal and less current in the antenna.

There are various elements that are capable of changing the polarization of the radio waves traveling in the atmosphere. A change can occur from vertical to horizontal, or vice versa, or to a combination of vertical and horizontal, known as elliptical polarization, when the signals travels in the ionosphere. Due to this, the receiving antenna that is polarized in a specific way receives only a partial signal, based how the incoming signal and the antenna are aligned. No matter when the antenna is horizontally or vertically polarized, it will be able to a receive certain amount of the signal when the radio wave signal comes in from the ionosphere.

Antenna gain: The concentration of radio signals might not be the same in all the directions. Its concentration in a specific direction is known as gain. The gain of the antenna increases the strength of the signal in that particular direction. This concept is used in communicating in the preferred direction; the signal strength is thereby increased. The gain can be created by the antenna by transmitting radio waves that add to other waves in the preferred direction, canceling out the other direction. However, an important thing to keep in mind is that the gain only focuses all the power in one direction; it doesn't create the power.

Applying this logic to antennas that radiate equally in all the directions, known as isotropic antennas, they cannot have gains. In reality, isotropic antennas do not exist; they are only used in theory. But another type of antenna exists that radiates equally in all directions in horizontal space—this kind of antenna is known as omnidirectional. Omnidirectional antennas are used for communication purposes over a widespread region. The antennas that have their gain in one direction are known as directional antennas, or the beams, and they are used for communication in one desired direction. The gain of an antenna is measured in terms of decibels (dB) and is always in respect of reference antenna. For example, when the gain is measured with respect to the isotropic antenna, it is known as decibel i (dBi). Therefore, gain is never an absolute quantity, just like voltage; it is always measured against a reference antenna.

All these concepts come into the picture because the radio signals vary in their strengths intensely. Generally, at the input side of the receiver, the signal strength is less than one ten billionth of a watt. But the signal that is emitted by the transmitter has the signal strength in kilos. Therefore, it can be seen that electronic circuits, antennas, etc. change the strength of the signals by several units.

Radiation patterns: So far, we have seen that antennas radiate signals uniformly and that there are several factors that affect this distribution. The best and simplest way by which the distribution pattern of any antenna is described is through a graph, known as the radiation pattern. The radiation pattern for transmission and reception for an antenna is the same, which means it follows the same distribution pattern for sending, as well as receiving, signals.

The most common type of pattern is the azimuthal pattern, which has to do with the gain of an antenna in the horizontal direction of the antenna. It can be created by imagining that one is looking down to the antenna.

And then there is the elevation pattern that describes the distribution of a signal by an antenna in its vertical direction. It can be imagined as looking at the antenna from the side. The radiation pattern of an antenna might vary with the change in frequency, as a change in frequency changes the wavelength of the signal; however, the dimensions of the antenna remain the same, causing it to act differently.

Once we have the radiation pattern of an antenna, we can see which region has the greatest antenna gain. This is known as the main lobe of the antenna, and all the other regions of lower antenna gain are termed as side lobes. To understand how useful an antenna is in rejecting the interference or noise from its surroundings, two important ratios are considered—front-to-back ratio and front-to-side. While front-to-back ratio is the antenna gain in the front, or preferred direction, the front-to-side ratio is the gain in the front, at the right angles of the antenna.

Chapter 5—Transmission and Reception of Radio Signals

We've discussed the basic equipment and propagation mechanism that hams use, along with basic electronic components and concepts around these components. So we are all set to dive deeper and learn about the real amateur radio, where the dials are turned, the knobs are rotated and meters go crazy oscillating back and forth.

The Band, the Frequency and the Mode

It doesn't matter whether you are using separate equipment for transmitting and receiving, or using one transceiver; two important functions that you need to know about are the mode and frequency. Amateur radio works a little differently from other radios in the way that it can be tuned anywhere in the defined band. Ham ram operators can only use channels preassigned by authorities in a provided 60-meter amateur band.

Amateurs are also allowed to use different kinds of modes, unlike various other radio services that can use only a single mode. Although amateurs are given flexibility over various things, they are expected to know how to select and set a frequency and choose a mode. The onus of selecting the frequency, the band and the mode is on the operator, and therefore he must know how to do it.

The first thing is to select the band, if you have the authority to tune your radio to different bands. Once the desired band is selected, next comes the frequency. The process of choosing a frequency to operate your radio on, within the selected band, is known as tuning. The controls that help you tune your equipment are known as variable frequency oscillators or the VFO. There is a difference in the way VFOs today work, versus older radios. Once, the VFO was knob used to alter the resonant frequency of the circuit, which in turn was used to determine the operating frequency of the radio. Conversely, the VFO controls in modern radios are controlled by microprocessors that automatically alter the frequency. Selecting the right band and frequency is important because one cannot use the amateur radios on non-ham frequencies. Since the instrument won't be able to read or hear outside the defined bands, you cannot transmit on any band other than the MARS frequencies for security reasons, expect for certain emergency cases.

Next is selecting the mode. If you have radios that can operate on multiple modes, you need to select the mode for your signal. Available modes include FM, AM, Morse code and Data/Digital. Radios that operate on higher frequencies are all multimode. On the other hand, most handheld radios that are also used as repeaters are single-mode radios that operate only on FM.

For easy access, the set frequency and mode are stored in memory channels so that they can be easily recalled at later stage. These channels are provided solely for the purpose of quickly tuning the radio to the desired frequency and mode. Some radios have memory channels that can also store information on the power level. All this enables easy access to frequencies and repeaters.

Functions of the Transmitter

While there are radios that operate on modes such as AM and CW, it is the RF power control that controls the transmitted output power of the radio. Another control that impacts the output power of the transmitter is the microphone gain, which controls the speech audio level applied to the modular circuit of transmitter. For transmitters that operate on AM and SSB mode, more modulation means powerful signal sidebands and a strong signal at the reception. However, too much speech audio can harm the signal by distorting and interfering with others.

Another control that is often found in radios is the speech processor control. This control enhances the power of a transmitted signal by amplifying the weak parts of the signal. While this process can cause the signal to distort a bit, it makes the reception better, especially in situations when there is too much noise in the surroundings. However, processors and compressors must be used with care as processing or compressing the signal can cause it to distort and may even cause interference with the surrounding frequencies. To avoid the interference with nearby frequencies when you are measuring the output of your transmitter or adjusting it, you can use a dummy load, which is a heavy load resistor that is capable of absorbing/dissipating the power from the transmitter.

The transmitter output power is measured in terms of peak envelope power (PEP). This is the power when speech content input into the microphone is the loudest.

We just read that processing and compressing causes distortion of the signal. Even modulation, when done in excess, can lead to distortion of the transmitted signal or interference with the surrounding frequencies. These unwanted outputs are known as splatters. When an FM signal is overmodulated, it causes excessive deviation, which can cause interference on the nearby channels. Sometimes FM transmitters are adjusted wrongly and deviate from the normal speech levels. This can be caused by speaking loudly in the microphones. Overmodulation can also occur due to the RF feedback, which means the signal from a transceiver is picked up by another microphone circuit that distorts it. Overmodulation of an AM signal possibly results in distortion of the transmitted signal.

Overmodulation can be prevented by use of an automatic level control (ALC) circuit that automatically reduces the output of the transmitter as the maximum limit approaches. This circuit also helps in keeping the signal clean of distortions.

Overmodulation of a CW signal can lead to interference as it generates the key clicks, which are the brief, sharp sound heard on a frequency near the distorted signal. These clicks are created by a transmitter circuit that tunes its signal too frequently, such that it causes dots and dashes. Therefore, the way of controlling the clicks is by reducing the adjustments in the setting of the transmitter that impacts the way the CW signal is created.

Functions of a Receiver

The function of a receiver is tough; it has to pick out only the required signal from several signals that are available for reception. However, the receiver circuits available today are quite effective, as long as they are adjusted the right way to pick up the right signal. So it is important to understand how a receiver works.

One of the key controls in a receiver is the volume control or the AF gain. This is the control that sets the listening level of the device. On the high- frequency band, we have RF gain, which adjusts the sensitivity of the receiver towards the signal that's coming its way. If it is seen that a heavy load is coming towards the receiver, its signal strength is reduced by using attenuator, which is used whenever any excessively strong signal approaches the receiver. Therefore, with the help of an attenuator and the RF gain, a lot of distortion can be filtered out when the band is full of heavy load signals.

Then there is automatic gain control (AGC), that automatically adjusts the sensitivity of the receiver towards the output signal so that it is relatively constant for strong, as well as weak, signals. A fast and effective AGC is needed for CW signals and data, whereas the AM signal can manage with the slow response AGC. For FM receivers, there is no AGC needed.

What does the receiver receive when there is no signal present? In order to answer that question, a special kind of circuit was invented. This circuit, known as the squelch, prevents the receiver from listening to noise in cases where no signal is present. The squelch turns the audio output off for the receiver when there is no signal available. Its threshold can be controlled using the squelch control, which is the level at which the mute is turned off for the output. This means that if the output is not muted, the squelch is working.

The receiver is also made to reject all the unwanted signals with the help of band-pass filters. The receivers are comprised of narrow filters so that they can pass through only select frequencies. So they have these band-pass filters right at the input that pass the signals only from the required band. Furthermore, the signal has to pass through IF, or the intermediate frequency filters that are narrower to reject signals from the select band. There are several types of filters, based on the specific requirement, so there are wide filters that are used for receiving the phone signals, and then there are narrow IF filters that filter only a few signals, for example for data mode and Morse code. Having multiple filters on the receiving end helps in rejecting interference or noise and, based on what is desired at the receiving side, the filters are selected with the right bandwidth.

Receivers are equipped with another type of weapon to filter out all unwanted noise and interference—notch filters. These are high-efficiency filters that remove a narrow range of frequencies from the output of the receiver. They are particularly helpful when high-quality output is needed out of an interfering tone. The notch filter eliminates the undesired tone, leaving the desired output that can be in the form of code, data or speech. Receivers can also have noise reduction that eliminates all audio noise or interference with the help of digital signal processing (DSP). DSP is a method that involves use of a special type of microprocessor in radio to perform filtering and other operations on the signal received. Some of the functions performed by DSP are variable signal filtering, noise reduction, audio response tailoring and automatic notch filtering. DSP works on the received signal after converting it into digital form. The microprocessor then performs the desired functions of filtering, etc. on this digital signal. Finally, the filtered digital signal is converted back into the audio signal (or into its original form) and fed into the headphones or the speakers. DSP has emerged rapidly in the modern radio.

Another function that is often used at the receiving end of the receiver is receiver incremental tuning (RIT), which is a fine-tuning control often used in reception of SSB signals. This type of tuning allows the hams to tune into a station that is little off the desired frequency or it helps adjust the pitch of voice of operator.

Radio spectrum is full of different kinds of noise and therefore, various types of special circuits are used to eliminate these noises or at least minimize their impact. A noise blanket is a weapon that senses the sharp pulses from motors, power lines or vehicle ignition systems, and accordingly mutes the receive upon detecting these pulses. Then there is a noise limiter that clamps that signal at a desired level, instead of removing the noise from it. These seem to be particularly helpful when the signal is at lower frequencies and there are nearby storms creating static crashes.

Receivers usually have the meters attached to them to indicate their strength with the help of a moving needle or a bar at the bottom of their display. They show the change in strength in terms of S units that are numbered from S-1 to S-9, with S-9 being the highest number in terms of signal strength. Whenever there is a change detected by one S unit, it means the strength of the signal has changed by a factor of four or so. When the strength of the signal is reported to be more than S-9, it is measured in decibels above the level S-9. For example, 10 dB over S-9.

Digital Form of Communication in Amateur Radio

With the emerging technology where everything in this world is being digitized, amateurs are not left behind. They have also developed techniques for communicating signals in digital form. Today, in the world of amateur radio, there are digital techniques that are deployed for exchanging data in digital form by converting ones and zeros into signals that are at the same frequency as a human voice. The radios are so designed that they transmit and receive these signals as either FM or AM signals. Data modes, or digital mode, combine modulation, which is when the information is added to the radio signal, with a protocol that is comprised of the rules employed in packaging and exchanging the data. These protocols also manipulate the way the receiver and transmitter coordinate to exchange the desired data.

There are many reasons why digital modes are being used for communication purposes. If voice and CW signals are used, they are not capable of rectifying the errors caused due to interference or unwanted noise in the surroundings. However, the special codes used in digital communication allow the receiver to detect and correct these errors. The resulting signal is much better in quality and is error-free. However, amateurs are constantly trying to improve their protocols so that they can further advance the quality of communication.

Digital mode also uses modems or modulation-demodulation, which changes the audio signal to the data signals. These are the devices included in the radio circuitry to perform the desired conversion. Sometimes even a microprocessor is included, along with the modem, to perform the rules defined by the protocol. When the two are combined, it is known as a terminal node controller or TNC. These days, different combinations of modulation methods, codes and protocols are used to create digital modes that can perform at their best in the desired range of frequencies. The way they are combined defines the performance and speed of the communication. Another thing that affects the radio communication is the frequency range—HF or UHF/VHF—as the propagation is impacted by the change in frequency.

We just discussed how data is changed into tones for digital communication. However, this is not enough for reliable communication. A signal experiences various kind of disruptions when it propagates from transmitter to the receiver. There are noises, interferences and fading elements that often create errors in the data stream. If precautions and preventive measures are not taken, the received information will have all these errors in it. Therefore, it is important to incorporate some kind of mechanism for removing or detecting these errors in the output signal. To achieve this, some codes are included in the data elements to let the receiver know there is an error. For example, some of the codes that reach the receiver come with an extra parity bit that indicates an error in a single character of the received data. Similarly, there are other protocols that add special characters or data to the signal so that the errors can be detected by the receiver and sometimes even be corrected.

For signals at higher frequencies, the most common digital mode is the packet radio. Packets are tiny entities that are transmitted in bursts. To transmit individual characters as radio tones that are alternating rapidly, frequency shift keying (FSK) is used. Then a receiving modem and TNC reassemble this data into a proper sequence once it is received from the packets. In a packet, the characters are transmitted at a frequency of 1200 or 9600 baud or so, giving an overall throughput of about 400 or 3000 bits per second for a packet system. Each packet is comprised of a header and the data. The header of a packet comprises all the information related to the call sign of the receiving station and the checksum to let the receiver know about the errors in the data signal. Upon receiving an error, the receiver requests the packet be retransmitted and keeps asking for it until the correct data is received. This is known as the automatic repeat request, or the ARQ. This way the receiver ensures that the data received is error-free.

To connect to stations that do not fall in the range of stations, hams can use relays known as digipeaters. These relays store the data in packets and retransmit it upon request. These packets are then sent by individual stations, or nodes, and therefore, if the operators have information about the packet stations and their locations, packets can be forwarded from one station to another, even if they're far away.

The digital modes that are designed for person-to-person communication in real-time are known as keyboard-to-keyboard modes. These modes are quite popular on the lower frequencies, such as HF, and are also found at frequencies that are right about the CW signals. One of the oldest keyboard-to-keyboard modes that was invented in the 1930s was radioteletype, using the Baudot code. This method used to utilize teleprinters to display the received characters but today these are generated by computer software.

Another amateur radio that's gaining much popularity for public communication is the Winklink System that uses airwaves to send emails. It combines digital mode with an email program and pocket radio. Hams who are always on a move use a gateway known as a radio message server (RMS), which is connected to email services via the internet. When a person wants to connect to RMS while on a move, the emails that need a server connection are retrieved, irrespective of the current location of the ham. This way the hams connect to an RMS station even when they are at sea, thousands of miles from land.

When digital mode is set up for communication, one must be aware of how a station can be configured. First, a TNC must be connected to one of the digital ports of the computer via a COM port or a USB interface. This TNC should then be connected to the microphone input of the radio and headphone or speaker output. In case a sound card is used, it must be connected to the microphone input of the radio and the speaker or the output of headphone should be connected to the input of sound card. If a sound card is being used, there is also a need to use digital communications interface to isolate the computer from the radio to prevent RF feedback or noise from impacting the data signal. To correctly transmit and receive the digital signal, ensure the following elements are properly configured:

- **Digital interface:** If the computer is being used in the setup, it should be connected to the data interface to ensure control signals are properly connected.
- **Audio level of the transmitter:** Caution must be exercised so that the voice is not overmodulated due to excessive signal strength at the input of transmitter.
- **Audio level of the receiver:** The receiver output must be at an optimum level so that data interference can easily be converted back into data.
- **Transmitter control:** While tuning the transmitter, there might be a need to connect it to the input of the radio.

Another important component of digital mode is the internet gateway, a special kind of digital station that is used to connect to the internet for all the data sent from other amateur radio stations. Most internet gateways are set up to just forward the messages, such as the Winklink station. These systems receive and send messages with a recognized internet email address. There is another type of gateway that is used so that a computer can connect to the internet. The radio is equipped with an inbuilt interface with the Ethernet connection to which the computer can be connected.

Ensure all rules and regulations related to business and commercial messages are applicable to internet gateways. For example, one can certainly not send emails to employers using this gateway. And since much of the internet is filled with business and commercial activities, extra precautions must be taken to ensure the rules are followed.

Use of Batteries and Power Supplies in Amateur Radio Services

There is no doubt that proper transmission and reception of radio signal is important in amateur radio, but having a solid power source is also equally important. If we have a proper power source, it ensures a noise-free, clean, good quality radio signal. How? Well, an inefficient power source whose voltage is either too high or too low, or which supplies a current that is not enough for the equipment, can totally damage the complete circuit or equipment. So let us look at different kinds of power supplies that are normally used in radio equipment.

For radios, the supplies that operate from household AC power are most commonly used. For a radio to be able to use this power, the AC power is first converted into a smooth DC current. A power supply has two main units—the output voltage and the current that it can supply. That's why, whenever a power supply is rated, these two units are mentioned, for example—a 12volt, 20-amp supply. The current rating of the power supply should be at least as much as the sum of total that's needed for everything.

However, whenever there is a need of more power, especially in cases of disaster or public services, AC power might not be available and therefore, inverters and generators are needed at that time. A generator is a device that converts the mechanical energy of an engine into AC or DC power. On the other hand, an inverter converts DC power into AC. Inverters basically have DC batteries, the power of which are converted into AC. So the AC power output from either generator or invertor can be used to run devices or charge batteries. These devices are rated in watts, which is the measure of the power that they can provide while maintaining the output in a specified voltage range. Also, if the inverter produces a distorted output, the electronic equipment that is going to use the power can get damaged or overheated. So it is recommended to use inverters that give out a sine wave output.

Another important factor is voltage regulation. The voltage output of a generator and invertor vary a lot more than voltage given by other sources such as an AC utility grid. The generators that are made to run motors and tools are comparatively less expensive, but for radios, better voltage is more important than the cost for the type of functions they are involved in.

Another source of power is batteries. Unlike generators and inverters, batteries provide the DC power output that can be used by portable radios and emergency power devices when using AC power is not practically possible. Batteries are comprised of one or more cell, which is an entity that contains chemicals and produces a current through chemical reactions. The cell is created in a way that the chemical reaction doesn't occur in the path of electrons between the cell terminals. But when all the contained chemical is exhausted and the reaction doesn't happen anymore, even the current doesn't flow in the batteries. Cell voltage is determined by the type of reaction that takes place in it. The cells in the battery are connected in a series in order to ensure the voltage from all the cells adds up together. For example, six small 1.5-volt cells are joined in a series to make a nine-volt battery. The 12-volt battery that is generally used in automobiles is comprised of six 2-volt cells.

There are different types of batteries, but they basically fall into three basic groups:

- **Primary or disposable:** These are batteries in which the chemicals react just once, and when the chemical is used, it cannot be used again.
- **Secondary or rechargeable:** These are batteries in which a chemical reaction can happen multiple times, so the battery can be recharged.
- **Storage batteries**: These are rechargeable batteries that store energy for the long term.

If you want batteries to last longer, try to minimize the amount of current drawn from them. If a battery discharges at a low rate, it remains cool from inside and minimizes the losses due to the internal resistance of the battery. So, to get the best out of batteries, they must be stored in dry and cool place. Sometimes, batteries are even stored in the refrigerator—not in the freezer— to keep them cool. When exposed to heat, the chemical stored in a battery tends to self-discharge, impacting the battery's efficiency. And if there is moisture, it will cause the charge inside the battery to leak between the external terminals of the battery. Therefore, it is recommended that batteries be regularly inspected for any leakage or damages and undergo regular maintenance.

Care must also be taken while charging batteries because not all chargers are meant to charge all kinds of batteries. Each battery is unique in its composition and size, and hence requires a unique method of charging. While overcharging batteries can damage them, charging them too little can keep them from providing their best output.

During emergencies, storage batteries are often used for power sources in place of AC power. Since they have a huge capacity and can store a good amount of power, they can be left connected to a charger that charges them with a small current. This process is known as float charging. Ensure that, while charging, the charger switches to this lower amount automatically, or it can totally ruin the batteries of the device by overcharging. When there is no commercial power available, a 12V storage battery can be recharged with the help of a vehicle's battery. However, ensure the system doesn't get overcharged.

Storage batteries have a huge capacity to hold energy and must be treated with care as they contain strong acids which cannot be allowed to leak. They also tend to release flammable gas, and if this happens it can lead to an explosion. Therefore, they must be stored in a cool, well-ventilated place and only be carefully used for charging devices y. Any mistake or short circuiting due to faulty wiring can lead to hazardous results. Once the battery is used, it should be thrown away. Batteries that are toxic should be considered for a battery recycle program.

Handheld radios use battery packs which are groups of individual cells that are connected together in order to create a battery of higher voltage so that the output from the radio is more powerful. Most battery packs are comprised of rechargeable cells. This means the packages are made up of several individual batteries that are rechargeable, connected together in a way that the package is considered a single rechargeable battery. The internal batteries that make up these packages are made up of different types of material, based on which the battery pack is categorized. The cells that are used can be nickel-metal hydride (NiMH), nickel cadmium (NiCd), etc. and some of these cells are used only in the packs; they are not sold as separate cells due to special charging requirements. For a given battery pack of same size, if we compare the capacity of these different material, lithium-ion (Li-ion) comes with the highest capacity, followed by NiMH and NiCd. Higher capacity means the battery will last longer. Even the disposable alkaline batteries that are available at any store have the same energy capacity as that of a nickel-metal hydride (NiMH).

Rechargeable batteries are preferable over disposable batteries because of two main reasons—they are convenient to use and are comparatively less expensive. The radios that come with these rechargeable battery packs have a charger. However, these days, fast chargers are also used to charge a pack in less time, even acting as a convenient stand for the radio. However, in case of emergency, conventional disposable batteries are preferred because they do not have to depend on a charge for their energy, and that's the reason many types of radio come with a battery pack of disposable batteries, although the transmitted power might not be that great. Particularly if the radio is being used for public service or disaster communication, ensure it has the option of using disposable batteries.

Radio Frequency Interference (RFI)

With so many electronic and electric devices being used today, interference between frequencies is bound to happen and that's why we see interference between the radio and these various types of devices that also operate on radio frequencies. This interference is known as radio frequency interference or RFI. RFI can be experienced either in the signals coming in or out of amateur radio equipment. The higher the power, or the closer the devices, the more intense the interference. So let's see what happens when interference happens, what can be expected and how it can be prevented or eliminated from our signals.

Filters are designed to both prevent unwanted signals from being transmitted and prevent unwanted signals from being received. However, the nature of interfering signals must be known to understand what type of filter will best do the job. Some filters that are used to filter out interference include low-pass filters, high-pass filters, low-pass transmitting filters and so on. These are used in a variety of ways to reduce the RFI. Answering machine or telephone interference is often eliminated by placing a ferrite core on a power supply cable, along with a low-pass filter on a phone line. The two pieces of equipment prevent the interfering signals from getting to the phone device. In the case of TV, a high-pass filter is used in the antenna feed line and ferrite core is used on the power cable.

The high-pass filter prevents the strong amateur signals from being received by the TV receiver and causing an overload. A ferrite core can eliminate the interfering signal from entering the TV. A low-pass filter, if placed on the output of a transceiver, keeps weak harmonics from entering the TV or radio. In certain cases, where there is a need of an additional effective filter, a notch filter can be tuned in to prevent the interfering signal.

AC power line filters are known as low-pass filters that help in keeping RF signals at bay so that they do not pass into or out of the AC power connection. These low-pass filters tend to reject all signals that have frequencies higher than a few kHz. The high-pass as well as low-pass feed line filters that are attached to antenna feed lines help eliminate RF interfering signals that are below or above a desired frequency. Then there are common mode, or RF choke filters, that are made out of ferrite material, used to prevent the RF currents from flowing into unshielded wires, such as AC power cords, speaker cables, etc. These chokes are also used to reduce the RF current flowing outside the shielded microphone, computer cables and audio cables.

All this is okay, but a device doesn't have to be a receiver to experience the interference. Even other devices such as lamps, music players, etc. can experience interference and are often seen to be greatly affected by unwanted RF signals. And once the RF signal enters any electronic device, all the components of that device, including the ICs, transistors, etc. can turn it into current and voltage that affects the normal functioning of the device, distorting its output in most cases. This is known as direct detection. One can make out it has happened if there are pulses coming out from a transmitter whenever it is turned on or off. In the case of AM transmission, direct detection can lead to distorted voice, and strong FM signals will be heard as hums. So, to eliminate the interference caused due to direct detection, a RF signal must somehow be stopped from entering into the equipment.

One of the most effective materials for achieving this is ferrite core. As discussed earlier, it is a ceramic magnetic material that can suppress RF energy over a broad range of frequencies, such as VHF or HF. The material is available in different kinds of mixes to form a composition that can absorb unwanted signals. One well-known type of ferrite in this category is the snap-core, which is formed by mixing things to the actual rectangular block of ferrite that has a large hole in it, sawn into halves. The two halves are held together with the help of a plastic case and this allows the cables to be wound on the core, irrespective of the fact that they already have the connectors attached to them, such as video cables. Generally, ferrites are available in the form of beads or round cores. Then there are wires that are wound around or passed through these ferrites. The beads that are mentioned here are large enough to slide over a coaxial cable and are then secured with the help of tape. This ferrite core, having a wire wound over it, forms a type of filter known as a choke.

Another type of phenomenon that impacts the working of a receiver is overload. An overload is caused when strong signals impact the ability of the receiver to reject them, and this type of overload is known as fundamental overload. Generally, overload causes serious interference to various channels of TV or radio, and then the equipment consuming these strong signals often fails to reject them if they fall outside their receiving bands. In a similar fashion, an amateur radio might experience noise across its band in case a strong signal is detected in the band. If this is attenuated either by removing the antenna or by using an attenuator at the receiver, the interference might disappear if it was due to overload.

A high-pass feed line filter can be used with TV and FM receivers at their antenna input, to reject strong signals at lower frequencies from amateur HF stations. If a filter is attached at the transmitter end of the antenna, it will not solve the issues caused due to overload, since the problem lies in the reception. It is the responsibility of the owner of the receiver to look into the overload issues caused due to signals coming from an improperly working transmitter. However, if the frequency of the interfering signal is too close to the frequency of the desired signal, a high-pass or low-pass filter cannot come in to rescue as it would attenuate even the desired signal. Say, for an instance, if a TV receiver is overloaded by a two-meter transmitter, a band reject or notch filter is needed to attenuate the two-meter signal without even filtering out the broadcast channels.

Amateur, as well as consumer receivers, experience overload from nearby stations. Although broadcast-reject filters can be used to attenuate these unwanted signals, the type of signals that are to be rejected must be known at the time of selecting the right type of filter, given that stations transmit on different frequencies.

In addition to overload, the output signals also contain some other undesirable elements due to minor imperfections. These are weak harmonics and other spurious emissions that can interfere with the desired signals of nearby equipment. Sometimes, strong interfering signals are generated if the transmitter that is being used is defective or misadjusted. To prevent radiating these harmonics from your station, the antenna feed line must have a low-pass or high-pass filter installed at the connection point of the transmitter. Therefore, as a matter of good practice, amateur stations operating at HF often use these filters to keep the harmonics generated by transmitter at VHF from reaching the antenna. Even if the transmitter works adhering to all the FCC's rules, a nearby TV receiver can still pick up these harmonics, resulting in interference. The worst part about these harmonics is they cannot be filtered out at the receiver end since they operate on the same frequency as that of the desired signal. This kind of interference is known as in-band interference. Another thing to note here is that the filters that are to be installed in a feed line carrying the signal should be made to carry the full output power.

Interference in amateur stations is generally not caused by a transmitter. Typically, interfering noises from unintentional radiators are received by the amateurs. These are either caused due to leakage in electric circuits or generated as a byproduct from equipment. Some of the most common sources of these kinds of noises are:

- Thermostats, neon signs, motors and electric fences that release electrical arcs and generate noises that spread across a wide range of frequencies. The noise is strongest on the lower radio frequency bands and becomes weaker with increase in frequency. The AC power line filters do not always work on the offending equipment; however, the on and off pattern that is seen in the noise often provides a clue of its underlying cause. Say, for example, that the noise generated by an electric motor is there for a few minutes when it starts and then stops later.
- Loose connections, bare wires rubbing against each other, bad insulating elements—all these can lead to noises. If you try to drill down into these noises to find the underlying cause by using a broadband receiver, write down the identification number of the pole and inform the power company in your area. Do not try to touch or shake the pole as there might be live

wires that can fall right on you. Ensure you maintain a distance from these poles.

- Switching power supplies and consumer electronics generate noise at mostly HF. Once the source is identified, a power-line filter can be installed to eliminate the noise.
- The DC power system inside our vehicles can also be a cause of a type of noise, known as an alternator whine. It is audible with the receiver audio and also by others as a high-pitched noise. It can be eliminated by a DC power filter on the radio.
- Motor vehicles also create a noise known as ignition noise, which is a whine that varies with the speed of the engine. It is normally only heard when a vehicle travels past you.

There are times when amateurs experience interference from a random or amateur transmitter that is transmitting on an amateur frequency by mistake. Intentional interferences are rare to see. Whatever the case is, it is important to identify the source of this interference so that it can be eliminated. This is generally achieved by radio direction finding, or RDF, which utilizes various directional antennas and maps to identify the offending transmitter on the amateur frequency.

So it can be seen that there are multiple types and sources of interference and noises that often interfere with amateur radio signals. Amateurs are quite used to it now. Irrespective of the underlying source of these noises and interference, much of the interference can be prevented or eliminated by ensuring you follow the right amateur practices for transmitting and filtering. Here are some guidelines that can prevent a lot of interference from occurring:

- Try to eliminate interference to your own appliance first. Ensuring you do not interfere with your own home appliances is a good start.
- Ensure your station is in a good working state and has appropriate filtering, grounding and good connections, especially for the RF signals.
- Ensure all the wires and cables that are being used are well shielded to prevent interference or coupling with surrounding unwanted signals. If shielding is missing from any of the wires or cables, connect the shield properly

- Ensure that there are no sources in your own state that causes interference, such as poor power supplies, old worn-out motors etc. All this not only helps make proper connections but also makes it easy to determine the noise in surroundings.

After doing all the analysis, if you discover that your signals are responsible for causing interference to a neighboring station, or vice versa, you can follow the following suggestions while also reaching out to the ARRL website for a detailed solution:

1. Always be ready to help in analyzing the nature of interference—is it due to harmonics or overload? What is it that is causing the interference? If you can determine the underlying cause, finding an effective solution should not be that difficult.
2. If you are sure it is your transmissions that are causing the problem, you might not think twice before diving deeper into what is causing it.
3. If you are sure it is the neighbor's device or equipment that is causing the interference, offer to help to understand the source of this problem. Such noises are often an indication of something serious, so try to prevent the hazards. Again, do not delay in finding the source in order to come up with the solutions
4. Politely and patiently explain to the neighbor that it is against FCC rules and regulations to use devices that cause harmful interference or radiation.

The RF Grounding Rules for the Amateurs

In amateur radio stations, it is essential to take into consideration RF grounding along with AC safety grounding. The type of grounding makes a lot of difference in the systems. RF is something that doesn't keep your equipment at ground potential. It refers to a phenomenon where all radio equipment is maintained at the same RF voltage. The feed lines of your station and the cables act as transmission signal antennas. These feed lines and cables are, in turn, connected to your devices and the connections between them. Therefore, when all the equipment is maintained at the same RF voltage, there will not be any current flowing between them. But when there is RF current flowing into the station, it can lead to distortions in the audio signal, causing computer equipment to function with errors. Such cases are more commonly seen when RF current flows in sensitive data or audio cables that interfere with the normal functioning of your radio station, just the way a strong signal transmitted by you gets picked up by an audio system in the surrounding area.

Since every RF station is different, its RF ground will also be unique. Below are some general guidelines around RF grounding:

- The RF ground rod is specially designed and should never be replaced with a properly wired AC safety ground, which is wired and bonded, keeping the local building codes in mind.
- All metal equipment enclosures should be bonded together to a common ground bus.
- The ground bus should be connected to a grounded pipe that has a wide, short conductor like a heavy solid wire or a strap. Solid straps are considered to be the best candidates as they have the lowest impedance towards the RF current.
- If the braid is removed from the coaxial cable, it shouldn't be used for RF grounding purpose as the cable jacket does not press the strands of the braid to make good contact. The braid taken from the cable also starts rusting if it is no longer protected by the jacket.
- All the wires, connections and straps should be kept as short as possible, as lengths longer than nine to ten feet are good candidates for antenna operating on higher frequencies. The ground rods become ineffective at higher frequencies, such as VHF, because of the length.
- In bad conditions, a piece of screen or flashing can be kept under the equipment and connected to the bus.
- If the shack is above the ground floor, the connection length from the equipment will be too much for it to act as an RF ground on any of the bands. Therefore, a ground bus should be used in such cases between the equipment, to keep it at the same RF voltage.
- A ground floor, or even a basement, might be able to use cold water pipes for grounding purposes as they travel below the ground for many feet to connect to the main pipe. However, if you choose to use a pipe for grounding, ensure it is directly routed out of the house. If it runs all over the house before entering the ground, it won't do much good. Also ensure

the pipe is made out of metal as plastic pipes cannot be used for grounding purposes.

Chapter 6—Finding Your Friends; Finding Other Hams

We all need a support system to help us through rough times. Similarly, hams need mentors to support them in case they need any help. Fortunately, there are thousands of Elmers, or mentors, in ham radio clubs around the globe.

Radio Clubs

The easiest and best ways to get in touch with other hams is through the radio clubs that have been around forever now. Initially these were just groups of like-minded people who love amateur radio but over time, these groups took the shape of clubs that grew in importance and size. These clubs are great resources for mentorship and guidance. You can find them in your local area by checking websites or on an ARRL site. Just enter the city, state and zip to locate your nearby clubs. The problem is not in finding clubs but in finding relevant clubs—choosing one that can be of help to you. Once you select the club, try to socialize and make contacts—as many as possible—and show up frequently to show and feel that you belong there. Try these things to help you socialize better:

- Whenever you get an opportunity, sign up for meetings.
- Introduce yourself to whoever you come across—it's a club, and people are there to interact and meet others.
- Show up at the club frequently and be a part of meetings.
- Wear your name tag or have any other identification paper that announces your name and call sign. This will help people remember you.
- Attend club parties and other activities.
- Look presentable; no one likes shabby-looking people.
- Introduce yourself to the bigshots of the club—identify yourself at the beginning of the meeting.

Once you are a regular at the club and people start knowing you, find out the details about the jobs listed below and see who is currently managing them. By contributing towards these events or activities, you can strengthen your bond with other hams.

Hamfests – Whenever there is any event, the host needs help for organizing it. Therefore, any sort of management expertise or social networking skills would help.

Club stations – If your club has its own repeater station, some sort of maintenance work is always needed, such as changing batteries, cleaning, testing the radio and so on. Connect with the station manager and ask how you can be of some help. You can also use this opportunity to become familiar with the equipment, which will help you in your amateur radio career.

Field days – Hams have operating events and they need help in planning and organizing them. See if you can help them with arranging food, tents, etc.

The Relay League of America

The American Radio Relay League (ARRL) is known as the world's oldest functioning amateur radio organization. The ARRL was founded with an aim to provide various kinds of services to hams around the world and plays an important role in representing the ham radio cause to the government and public. ARRL is a leadership organization that has been the strong support of hams for almost 100 years now. Therefore, it is good to have an understanding of this large presence that offers some valuable services to hams.

The most obvious benefit of ARRL is that it publishes *QST* magazine every month, which is the oldest and largest ham radio magazine, comprised of articles and reports on both technical and non-technical topics, regulatory information around the topic, etc. Most of the biggies in the ham radio equipment industry advertise in *QST*.

In addition to this printable magazine, ARRL also features:

- Numerous emails, newsletters and bulletin
- Daily news and interesting stories
- An active ham radio shop that is open 24/7
- The technical information service, which is an extensive reference service comprised of several technical articles

The ARRL field organization helps thousands of volunteers, spread across 80 different sections, in 15 divisions, coordinate. While the league is headquartered in Newington, the ARRL is largely a volunteer organization. The ARRL field organization also consists of the world's most extensive non-governmental radio network, the National Traffic System (NTS), which responds to all types of emergency situations and is active all the time, sending radio messages to different parts of the world.

In addition to all these organizational and administrative functions, the ARRL is also the largest sponsor of operating ham activities. These sponsors fund various competitive events, technical and emergency exercises and award programs.

On the ham bands, the most visible aspect of the ARRL is its headquartered station, W1AW, which is a worldwide beacon offering Morse code practice session and bulletins to hams across the world. The ARRL also provides the volunteer examiner (VEC) service, helping thousands of new hams in getting their licenses, obtaining call signs, renewing their licenses and updating the license information. Generally, the FCC maintains the staff to administer these programs, but when it fails to do so, the ARRL and other organizations come forward and create VEC programs that are necessary for the health of today's ham radio.

The amateur radio service to regulatory bodies and government is one of the ARRL's most important functions. Today, in the telecommunication driven world, radio spectrum is one of the most important territories. Various other services want to get access to it, irrespective of the long-term consequences. The ARRL is also a trusted voice to Congress, as well as the FCC, that helps legislators and regulators understand the special needs and nature of ham service.

So you can see that the ARRL helps regulatory bodies and hams in various respects. Now, let us see how it benefits the public:

- The ARRL publishes a handbook which is a trusted and acknowledged telecommunication industry reference used by amateurs and professionals. Several technical references and guides are published, which include standards, conference details, etc.
- The ARRL facilitates emergency communication development. The ARRL, along with the Amateur Radio Emergency Service (ARES) teams, spends thousands of hours in public service every year. Although individual

amateurs play a key role in times of emergency, these organizations join hands with them to multiply their efficiency.
- The ARRL not only promotes ham radio but also spreads technical awareness. The league promotes and supports several programs for elementary secondary schools so that they can learn about ham radio service.

Ham Clubs, Specialty Organizations and Online Communities

Ham radio is a vast, deep domain comprised of many communities that fill our airwaves with different types of activities. However, the clubs and specialty organizations focus on the aspect of ham radio that deals with certain types of operations or technologies. Some of these clubs also have an international membership. You can easily find a specialty club around you by entering your area into a search engine. along with the word 'club.' You will see there are many clubs you can reach out to. While some of these focus on certain specific operating interests, other emphasize operating on a specific single band. The clubs also sponsor many contests and activities every year and give awards to deserving candidates. Some examples of these clubs include the Six Meter International Radio Klub (SMIRK), that promotes activities on the six-meter band on the amateur frequencies. The band has a special signal propagation. Another type of specialty club is the contest club. Hams really enjoy competing with other hams on-the-air and these events are known as radio sports. The club challenges interested hams to organize events and sponsor awards to encourage them to participate. This way the club builds awareness around techniques and stations.

Then there are long-distance communication specialists, the contest operators, known as DXers, who specialize in making connections with places that are off the beaten track. The desire to work with them results in learning about obscure sites when hams begin to transmit from an island far away. DXers form these clubs so that they can share their experiences and host other hams, building international communication along the way. Since DXing is of an international nature, these DX clubs have members from all around the world.

Another name in the world of amateur radio is that of TAPR, which is the Tucson Amateur Packet Radio. Instrumental in bringing the latest digital communication to hams, TAPR has been a part of digital communication development for several years now. The innovative communication technologies that TAPR members have created have been found to be useful in other industries also (not just ham radio), like pocket radio communication system technology, commonly used for public safety industries. TAPR is also working hard towards bringing modern digital communication technology, such as spread spectrum, to ham radio services.

Another class of organization that provides terrific communication opportunities to people who find themselves constrained by various physical limitations is Handi-Hams. Founded in 1967, Handi-Hams is a specialty organization created to provide tools that make ham radio accessible to hams with disabilities. The organization extends its support to people who try to turn their physical limitations into assets through the use of ham radio services. Handi-Hams also encourages all those who want to volunteer their services to Handi-Hams, including the international community.

Promoting ham radio to women, the Young Ladies' Radio League, YLRL, was founded in 1939. It is basically an organization that encourages women to become licensed hams, giving them all the help and support needed to work as on-the-air hams.

AMSAT, amateur radio, is another ham organization created to help coordinate launches of the satellites and oversee the development of its own satellite. Radio operations are a lot easier and effective when done through a satellite.

In addition to ham organizations and specialty clubs, hams also have online communities in which the members can discuss various aspects of ham radio services. These communities are available 24/7, offering support and resources to ham operations. They make hams stronger by adding structure to their operations and cementing their relationships. Let's now look at some of the online communities for hams:

Newsgroups: The USENET newgroups publish a lot of interesting information including ham radio services. Once you subscribe to a newsgroup, you can create a post of your choice as it encompasses a wide range of topics. However, due to this openness, newsgroups are generally unmoderated and have a higher amount of spam thanks to off-topic discussions. Nevertheless, you can always find good information in these groups.

Reflectors: Reflectors were the first type of online communities created for hams. These were actually email lists taken from one mailbox and re-broadcasting them. Using this method, the information is spread rapidly and effectively. For every ham interest, there is a reflector created. The new hams choose to join Elmer email lists, where they can find answers to their questions. These Elmer lists can be found online by simply searching for the keywords "ham radio elmer." Some reflectors offer most than just email distribution lists; they offer chat rooms, member management, file storage and so on. In addition to all this, reflectors are also an excellent way to find details about new techniques and equipment in the ham radio industry.

Portals: Portals are a comprehensive set of services that feature links to databases, news, reflectors, informative articles and various other services.

Ham Festivals and Conventions

Ham festivals, or hamfests, are one of the most entertaining and interesting events in the amateur industry. The mere thought of having an oriental market that showcases various types of technological artifacts, experiencing a perfect fusion of modern and archaic, computers, antennas and batteries makes those who love ham radios very enthusiastic. These fests feature flea markets where the sellers are from different background; some are commercial sellers and vendors and then there are treasures showcase from a tailgate. Hamfests are indoors, outdoors, big, small—all types.

There are also ham radio conventions that are a broader version of a hamfest. They have more variety in terms of activities and things that are displayed. These ham conventions can include commercial vendors, seminars, test sessions, etc. and often include meetings with individuals from other functions. Generally, ham conventions are based on an underlying theme, such as digital radio transmission and emergency operations.

If you want to find out about some of the best hamfests and conventions in the US, the best place to get all the information is the ARRL's website. When you browse through the link, you will find the hamfest link at the top of the home page. Click this link and it will take you to the Hamfest and Convention page. Here you can look for various hamfests and other details around them. Once you finalize which one to attend, you can buy the tickets online or at directly the venue.

When you are at the hamfest, you should take care of a few things, especially if you are new to this world of ham radio:

- ✓ If you are a novice ham, tag along with someone who has experience in ham radio services.
- ✓ Hamfests make a good market for radio accessories, so if you are looking to buy something, buy it from the hamfest as you can get if for a fraction of market price.
- ✓ Most of the vendors demonstrate how to use the equipment, so you can actually test the quality of the product before buying it.
- ✓ Don't hesitate to ask questions, if you have any. Hams are there for you and love listening to your queries. So, shoot!
- ✓ Try to avoid buying antique radio, unless you are very sure of what you are buying. They might have issues that can make your life difficult and you might end up struggling.

DXing

In addition to various different activities that hams are involved in, pushing their stations to make long distance contacts is one of the oldest activities in the history of ham radios. With the help of DXing (where 'D' stands for Distance and 'X' refers to the unknown), hams are able to make contacts over greater distances; making contacts with regular radio stations is not considered a DXing. The history of DXing goes back to the Marconi days, when Marconi started in radio by sending messages across long distances. There is always a station that is in desperate need to make contact with the world, and to help such a station is the purpose of DXing. In fact, there are thousands of hams out there, all around the world, who are doing a great job of making contact (known as QSOs) with someone far away. They are not bothered about the stations in their nearby location; they are always on a lookout to make contact with people located in exotic locations. If you get in touch with them, ask them about the details and geography about their location and you will see that most of them know not just the geography but also the history of that place.

It might sound strange, but the history of ham and ham radio has a strong connection with DXing. However, DXers shouldn't be confused with radio amateurs, who are also known as ham operators. Unlike hams, who have their own operating stations, DXers are not involved in the transmission of any information and therefore do not need any license. They don't even need to know about the elements, the circuitry and the electronics of the radio, but this doesn't make DXing any less challenging. The challenge lies in receiving signals that are incoming from an incredible distance. There are still untouched parts of the world which no one has ever heard about, but which could be connected with the help of DXers.

As receivers became stronger and transmitters became more powerful, the distances at which a station was able to make contact became a direct measure of quality. Hams eventually started exploring and playing around with different bands; they could tune in with the fluctuations with the ionosphere. And, with time, DXing equipment underwent many good changes. Today, hams are able to make contacts with hams across the country and across continents over HF as well as VHF/UHF frequencies. While the contacts on the HF frequencies are commonly seen, contacts on VHF/UHF, which are once thought impossible, are also quite common these days. One can send out an email around the world or log onto a chat room, but logging a QSO in the log or getting in touch with a distant station is something non-hams can never enjoy.

So let's get deeper into the topic and see how to do DXing on the shorter frequencies. These shorter bands are comprised of frequencies below 30MHz, on which the signals generally propagate all around the earth, bouncing between the two layers—the earth's surface and the ionosphere. The ease with which these signals propagate between the two distant continents on these shortwave bands impresses many. And since they propagate so widely, DXing at HF frequencies often happens across the world with stations that are trying to get in touch from different continents. However, it's not just the HF DXing that is thrilling; even the VHF/UHF DXing is quite exciting, although it calls for a different approach. So, the propagation is quite selective and involves a smaller number of signals being heard at one time.

All these differences make the contacting process quite different on shortwave band as compared to the VHF/UHF bands. At these frequencies, any contact that is made beyond the horizon is known as DX. Contacts are best made on CW or SSB, as the modes are quite efficient. However, they are short, as the band for long-distance propagation is open for only a limited time. Therefore, hams require a multimode transceiver to use digital, SSB or CW modes for DXing at these frequencies. When making this type of contact, one of the most important pieces of information that is shared is the grid square (discussed later in the chapter). While using a beam antenna, which you can point in various directions, gets you good results, a FM or dipole also work at certain times. With a beam antenna, you will have horizontal polarization, which is the norm for digital, SSB and VHF/UHF SSB. Other than the obvious benefit of DXing that it helps hams located in distant parts of the world to make contacts, it also helps hams hone their operating and technical skills. Pursuing long-distance contacts helps them learn various things about propagation, elements involved in propagation and the surroundings.

The skills of DXers are recognized by the ARRL and various other organizations, and they are awarded for their achievements. One such award is the DXCC, which is given for making contact with different countries. Then there is another award, the VUCC, which is given to VHF/UHF enthusiasts. In addition to all this, there are many other awards for hams. For example, some clubs give away awards for contacting their members; organizations offer awards for using low power (for example—the QRP ARCI certificate is given to those who make a contact that is performed at 1000 miles per watt of power). There are also radio contests that are organized to see who can make the highest number of short contacts in a given period of time. There are certain contests that are quite short, known as sprints, while there are some that extend over an entire week. Then there are contests that utilize one band while others span across multiple modes and various bands. If you come across a contest that is being held on the air, never lose the opportunity to make a few contacts. You will be asked for information, known as exchange, which includes details of your current location, a serial number (describing the number of contacts you have made in the contest) and a signal report. So, whenever you choose to participate in these contests, just ask what is it that you need to make available and the details will be provided to you by the contest station. Ensure you give just the minimum information that is required to identify you/your station, and let the exchange happen.

As you can see, pursuing these awards and participating in these contests can contribute a lot to making you a capable operator. The brilliant skills and stations possessed by DXers and operators are applicable to traffic handling and emergency operations as well. Various contest operators that are listed among the top performers also handle radiograms and participate in nets. These contests are sponsored by the ARRL and other organizations that run the entire spectrum of international events to encourage many quiet hams to participate. All the rules for these competitions or events are provided on the ARRL website.

One of the biggest amateur events of all times is the ARRL Field Day, which takes place every year on the fourth weekend of June. It is attended by North American hams, who head towards the fields, the parks, the hills and the backyards to celebrate. This event is organized to spread the awareness for emergency cases, and hence it includes a preparedness exercise in which many hams participate. The basic idea is to organize and set up a portable station and make contact with as many hams as possible on various amateur bands. As part of this event, while most of the groups focus on emergency preparedness, others work on the competitive aspect for different points and the rest consider it to be an annual picnic, plus radio operation. No matter how it is treated, ARRL Field Day offers a brilliant way to see so many ham radios gathered in one place. All the information related to the event is available on the ARRL Field Day page of the website.

Another type of contest, a physical one, which is quite popular is fox hunting. As the name suggests, it is similar to a fox hunt—hams look for hidden transmitters. For this, hams get proper training so that they can locate lost people, downed aircrafts and so on. For the hunt, you don't need much in terms of equipment—all you need to get started is a portable radio that has an indicator to show the signal strength, or a portable directional antenna. For training purposes, one of the hams is asked to hide the transmitter so that the rest can walk, ride, bike, drive around the area, attempting to locate the hidden transmitter. Amidst all this, a new type of radio sport has gained good popularity in the US—radio direction finding. As part of this sport, which is generally organized as events, hams explore the outdoors with the help of a compass and maps. The sport is another form of a radio fox hunt that utilizes the orientation skills of the hams.

Making a Contact

A simple ham-to-ham contact is basically the exchanging of call signs, which, in simple words, means exchanging each other's basic identities. However, most contacts do a bit more than exchanging just the call signs. Hams have casual conversations with other hams, coordinate with each other for public events and so on. However, no matter what kind of contact it is—formal or casual—there are certain common elements that are a part of all the communications. So, let us look at these general principles.

Making radio contact is very different from the usual face-to-face conversation. When people talk over the radio, they do not see each other; all they hear is the words, which are communicated in a special way. For example: "My name is Coke—Charlie Oscar Kilo Echo." Normal people wonder why hams talk like that. Well, when communication happens over the radio, it is difficult to hear the letters apart, for example—T, B and D sound similar. Moreover, non-English-speaking people have their own way of pronouncing letters. Therefore, it's best to use phonetics. For this reason, a standardized phonetic alphabet was developed by the International Telecommunication Unit (ITU) so that operators of all languages could use it as a standard for exchanging information. For example: A for Alfa, B for Bravo, C for Charlie, D for Delta, E for Echo, F for Foxtrot and so on. Each word was chosen in a way that it could be easily understood over radio. All the hams are supposed to learn these phonetics and always remember them for communication purposes, even when they want to call out their call signs. All those hams who specialize in DX contacts use the names of cities and countries, which are not always heard properly over the radio. For example, Norway might sound like November. Therefore, these phonetics should always be used for proper communication to happen.

It's important to identify yourself regularly when on air. And your identification while you are on the air is your call sign, so always use it whenever you want your presence to be felt or want to participate in a conversation. Also, it is not enough to just identify yourself to the ham you are in contact with; all those who are listening to you should also know your identity. So give your call sign in the beginning and then during the conversation. And then identify yourself even before concluding the call. The FCC encourages all amateurs to use the phonetic alphabet when calling out their call signs so that people do not misinterpret them.

When on the radio, only one person can talk at a time; therefore, when you are done with your part, you must make it clear so that the other person can start. Always remember that others can't see you, so they are dependent on you to tell them when you are done. This is the reason phone operators say "over," so that the other party can know it's their turn. When the contact is completely over, the operators use the word "clear" to let others know that the frequency band is available for other communication to happen. These are procedural signals that are used for proper communication.

Always ensure you speak clearly and talk a little slower than how you normally do. Also, always remember to tell others when you are done transmitting, since there is no eye contact. When using the digital mode, make an effort to punctuate and spell correctly.

Hams generally use the same frequencies on a regular basis, for instance, a ham might regularly monitor the 442.50 repeater or use the frequency 7.5 MHz for his communications. However, no matter how much time a ham spends on this frequency, he doesn't have any exclusive rights on it; he doesn't own it. This rule holds good not only for individuals but also for groups that meet every day on a frequency, at the same time. If they find that another group or ham is using this frequency, they should be ready to move to another one. Although it might be little upsetting and difficult to find another frequency when you have a meeting, hams are generally quite understanding and accommodating when it comes to using the band.

Also, it is important to let the transmitting station know that you have well-captured the signal, once you receive it. This is important so that the transmitting signal can take care of the quality next time while transmitting. If the signal was not well-received, and conversely, if the signals were good enough, repetitions can be avoided. The information about how well it was received is known as the signal report. In case of digital, CW and SSB, the report is sent in the form of RST—readability, strength and tone. Each attribute is rated on a scale of varying numbers—readability is rated on the scale of 1 to 5, strength is rated on a scale of 1 to 9 and tone is rated on a scale of 1 to 9. Readability refers to how well the signal was interpreted, whereas the strength is the relative strength of the signal received. Tone is used only in the case of digital and CW transmissions, referring to the quality of the signal. In case the signal has a pure, clear tone, with no noise or distortion, the signal has a tone of 9.

Another measure is that of the power level, which tells us how much power the station is generating or running. The FCC has laid out clear rules stating that hams should use the minimum power that is required to make contacts. But this does not mean that you should reduce the power to a level that it can be barely heard. The intention behind setting this rule is so that others are also given enough power for their requirements. If one station eats away more power that what it needs, others might run out of power. This is more of a concern in the case of digital, SSB and CW modes. In general, if the signal is able to manage on its own, do not give it the support of an amplifier.

We now know about the signal reports that determine how the signal was received, and we also know about the power level required for the stations. So, let us now try to locate other stations—other ham stations. The method for identifying the location of other stations always varies and depends on the reference. However, one of the popular methods for identifying location is the Maidenhead Locator System, which is also known as grid squares (discussed earlier in the chapter). According to this system, the surface of the earth is divided into a system of rectangles based on longitude and latitude, where each rectangle is determined with the help of numbers and letters. It is generally a four-digit code comprised of two letters, followed by two numbers. This code uniquely identifies a square of one-degree latitude by two-degree longitude.

We've discussed some of the things that are important while making contacts—the reports, the phonetics, the power, etc., but the question here is—what do hams talk about? Well, people who have heard hams talking, often hear them discussing weather, sports, events and so on. So, yes, they do talk about everything and anything, and there are really very few topics that are restricted. However, obscene and indecent language is totally prohibited. In addition to bad words, they also try to stay away from using provocative subjects that tend to create strong feelings.

So, after getting a basic overview of how contacts can be made, let's get to the real work of making contacts.

The most important part of making contacts, also known as QSO, is listening, and this makes our ears the most important attribute of the station. The ears of hams are alert and on 24 hours a day so that they can learn all about people coming and going. And that's why they also have to tune the band or monitor in to discover who's out there, and why.

Repeater contacts: Repeaters are just like the social clubs—they are manners defined for everyone and when anyone joins the "club," they are supposed to adhere to these manners, such as:

- Keeping the transmissions short and crisp.
- Always listening to ensure the repeater is being used.
- Identifying the station correctly and legally, and identifying yourself in the beginning of conversation.
- Pausing in between the conversation is required in order to know if other stations are trying to break in.

Despite all this, understand that the fact that you are using a repeater in no way means that you are not responsible for operating. Being the control operator, you are solely responsible for the operations and the repeater owner cannot be blamed in case of a violation of FCC rules.

Repeaters normally send a strong signal over a frequency and therefore, it is not required to send a long transmission to get listeners tune to your frequency. Sending just a call sign will do the job and let others know you are available for a contact. Listening stations also sometimes add a word or two to make sure you know they want to make a contact. Hams are heard saying, "X1AW is monitoring" or "X1AW is waiting for a call," which means they want others to know that they are available for a contact through the repeater.

If you want to respond or start a conversation with a station whose call sign you know, just state the other ham's call sign, followed by your own call sign.

We've discussed the signal report—this is basic etiquette repeaters are supposed to follow. Since the repeater resends the signal, the signal report will help in understanding how well it is heard. Along with the rating, the following terms are often included:

- **Full quieting**: The transmitted signal is strong and there is no noise.

- **White noise**: The transmitted signal is not strong enough and there is some noise present.
- **Scratchy**: The transmitted signal is weak and the noise is almost as strong as the signal.
- **Picket fencing or mobile flutter**: The transmitted signal is rapidly fading.
- **Dropping out**: The transmitted signal is mostly audible but there are frequent periods of no signal.
- **Broken or breaking up**: There are short periods of audibility but the signal is mostly unreadable.

All these pointers give an idea of how the signal looks as the output of the repeater. They often add a short curtesy beep to the retransmitted signal if the signal disappears. This is used as the "over" signal for certain listening stations, which means they can start speaking. Certain radios seem to add their own courtesy beep; however, it is not common on those that have it enabled already. What happens when someone is using a repeater and by mistake you interrupt them? Well, if it happened by mistake, just apologize for the interruption by saying, "Sorry, <call sign>, clear" and then wait for them to complete, or use a different repeater if you cannot wait.

If you hear that your signal is strong but distorted, it means your station is slightly off frequency. This might happen in cases where a radio control key is bumped, due to which the transmitting frequency is changed a bit. If the control key is getting pressed by mistake on your station, you can try using the lock feature to disable unintentional key presses. Even if you speak too loudly, it might lead to distortion in the microphone. So either lower your volume or hold the microphone at a distance from your mouth. Sometimes, even bad quality or weak batteries lead to distortions in the audio. Having a weak battery pack may cause a signal that is strong for a few seconds in the beginning and then fades away. This means the batteries are not strong enough to sustain a transmission. Therefore, it is important that you have a strong and working battery pack with you.

Another scenario is when the signal that is received by the repeater is strong but somehow the receiver doesn't respond. Such scenarios happen when hams use repeaters as on-the-air meeting places and are not looking for making contacts all the time. Therefore, it is always good to wait until an existing contact finishes and calls for making other contacts. This works particularly well in cases where you want to talk about a topic that was just concluded by the other contact. This way you can participate in events or nets on repeaters and be a part of the group.

Digital, SSB and CW contacts: Single Sideband (SSB) is quite a famous mode of transmitting voice signals on the HF bands. The band is so named because of the way it is different from the other mode, AM. While in AM mode, two identical copies of the information are transmitted, known as sidebands; an SSB gives out only one copy of the voice signal. The signal transmitted by an SSB is more efficient and also saves space in the spectrum. Most signals on HF are single sidebands only, so you just need to decide if you want to go with the upper sideband or the lower sideband. In reality, the SSB signals spread across a narrow band, either above or below the carrier frequency of the radio. So, in most cases, the decision is made based on the frequency—for frequencies above 9 MHz on an HF band, the operation takes place on USB, whereas for frequencies below 9, it is on LSB.

If you tune into an AM signal which has both USB and LSB, you can tune into either of the two bands—upper or lower. As you tune into the AM carrier frequency, the noise gets lower in frequency, precisely between the two sidebands. And when it becomes so low that the voice cannot be heard at all, you can tune to any of the sidebands.

While using fixed channels, contacts do not use repeaters for these modes, and therefore, finding and letting another station know about the need to make contact is done in a different manner. In this case, the station has to make a call that is long enough so that some of the listening stations can tune in to this frequency and identify themselves by stating the call sign. Hams usually do this by calling CQ, which is a procedural signal for "I am calling the stations." The station that is ready to make a contact with this transmitting station will say CQ multiple times followed by their call sign. This is just like saying, "Hey, CQ, <call sign>, calling and waiting to hear back." On a digital or CW mode, it sounds like, "CQ CQ DE, <call sign>, <call sign>, <call sign>, K," where DE means from and K means end of the signal.

Before a listening radio station responds to the CQ, it should ensure three important points:

1. The frequency they are trying to respond to is the one they are authorized to use.
2. Listening carefully to ensure it is not already in use—for this, just listen to the frequency for 5–10 seconds and if you don't hear anything, it means it is available to use.
3. Before you start with the actual conversation, start by making a short transmission to ask if the frequency is being used. It might happen that the ongoing conversation has been paused, therefore it is always good to ensure you are not interrupting. Also, ensure to state your call sign.

In case of repeater contacts, if you accidentally interrupt another conversation, just apologize for interrupting, give your call sign and then wait or tune in to another repeater.

So, once you manage to join in successfully and initiate the conversation, you might find yourself being part of a friendly conversation known as a Ragchew. If you are sharing the frequency with others, it is known as a Roundtable conversation. If stations which are located in rare places want to get in on a conversation, they generally exchange call signs and send signal reports so that the other stations can contact them. These are certain types of activities that take place on the amateur radio.

Once again, since we cannot depend on visual cues with radio communication, hams use certain methods to control the flow of the conversation. One of the commonly used methods is using the word "over," which means end of the conversation. Other than this, procedural signal K is also used, particularly for digital or CW contacts. Hams know they are not supposed to start talking until the transmitting station completes or requests the listening station to start. They say "over" to let the other stations know they are done and are not just pausing or in the middle of a conversation. If you are familiar with the other stations and if your signal is also of good quality, the use of the word "over" can be relaxed; however, there is no harm in always using it for all types of conversations. All these are different methods to ensure a proper, interruption-free conversation. Even so, we all know that no method can be foolproof enough to guarantee no accidents happen at the time two stations are having a conversation.

Since a ham's transceiver does not have the capability to receive and transmit at the same time, there is no way to know if a simultaneous transmission is happening, which is known as doubling, until one side stops. In case of doubling, someone from the listening side has to state "you doubled" to let the stations know it happened. Someone signals, one of the two stations asks for a repeat and waits for the other station to respond.

In other type of scenario, when two stations are having a conversation and you want to break in for a reason, you need to first gain their attention to get an opportunity to speak. This is known as breaking in, and you must handle it with style. The best way known is to wait for one station to finish and then quickly state your call sign. Therefore, it is considered good to pause briefly when you are in contact with another station so that the breaking station can get a chance to transmit. And if you find someone is trying to break in, be generous to pause and ask the break-in station to speak, because it could be an emergency break-in.

Repeaters are known for their great coverage capabilities; however, they are not used for making a contact. The reason for this is their wide coverage capabilities and signal strength. Since only a defined number of repeaters can share a specific band in the given region, hams must ensure they use it properly and wisely. Hams easily get used to using repeaters, and once that happens, they do not like communicating over simple channels even though the station they are trying to make contact with is close by. If one can listen to the transmitting station directly, why use a repeater? Using simple communication is a lot simpler and less public as there is no need to retransmit the signal over entire region. Using repeaters make sense only when you are trying to make contact with a weak station, but if the weak station is close by, you will hear them better directly as compared to using the repeater in the process.

Repeater stations are generally located in high spots, such as high buildings, towers, hills, etc. The output from the receiver of the repeater is retransmitted on one or more frequencies at the same time and the strong output signal is received over a wide region. This process in which transmission happens on one frequency, and receiving happens on another, is known as duplex communication. Since the repeater output is heard by several amateurs, the rules and regulations it follows are different from digital, CW and SSB contacts.

The very first step in using repeaters is finding one. To find a repeater that you can use, you first need to look at the band plan—look for the segment that is filled with output frequencies of the repeater. You can either use the scanning function of the radio or tune through the specific part of the band, but this doesn't ensure you will find the repeaters as they do not transmit continuously.

Once you hear a repeater, you can identify it by listening to it for a while and then, if you hear an automated voice that states a call sign, followed by some other information, you can identify the location by looking for this call sign online—this is the repeater's ID. You might also receive the Morse code. This way you can find all the repeaters in the area.

There is uniform distribution of repeater frequencies, known as channels. So, once you find a repeater, you know what frequency you can use. This is a good way of finding the scanning functions of the radio. By scanning the output frequencies of the repeater, you can find all the active repeaters present in the region. Once you know about the active repeaters, you need three things to access it—the output frequency of the repeater transmitter, the input frequency of the repeater receiver and the frequency of any control tones.

Repeater's output frequency: The output frequency of the repeater is the frequency at which you hear the repeater's transmission as well as the frequency by which repeaters are listed.

Repeater's input frequency: The input frequency of the repeater is the frequency at which you must transmit, where the repeater receiver listens. It leads to total chaos if every repeater owner has a different set of input and output frequencies; therefore, hams have decided to use a fixed set for both—input and output. The difference between the two is known as the repeater's shift or offset. This offset or shift is almost the same for all repeaters.

If the input frequency of the repeater is higher than its output frequency, it is known as a positive shift, and if the output frequency is higher than the input frequency, it is a negative shift. However, instead of remembering the two frequencies, you can just use a standard shift that lets you remember just one—output frequency. And you can always switch between positive and negative shifts by using the offset or shift key on the radio. However, radios are generally configured to use the standard shift, termed as auto-repeat.

Repeater access tones: Most repeaters do not send a signal from receiver to transmitter for retransmission unless they get a signal—a special tone, which is one of the 38 available frequencies. This special tone tells the repeater that the signal is for its use and that it needs to retransmit it. The CTCSS tones are included in the signal to indicate to the repeater that it should relay the signal. But before it is passed through a repeater, there is a need to find out which of 38 tones it is. You can refer to the ARRL Repeater Directory to find the tone from the given list that also has the output frequency listed for the repeater.

Some radios can also learn the frequency of the tone from the signal, which is known as a tone scan. Radios that are capable of determining this can be configured in a way that they can set their own CTCSS tone to a frequency without any external intervention. This helps when someone is unfamiliar with the area.

In case of repeaters, a signal, although at the proper frequency of the repeater, is not meant to be received by the repeater. So what happens is—most repeaters are installed in the vicinity of various other repeaters, broadcast transmitters and other devices, which transmit powerful signals that interfere with each other to create a false signal, known as an intermod, or intermodulation. These false signals appear on the input frequency of the repeater and get retransmitted, interfering with normal communications that are happening on that repeater/frequency. However, to prevent these false signals from getting picked up by the repeater, the repeater receiver should listen to the tone in the signal. If there is an improper tone detected, it means the signal is not meant for the repeater and hence shouldn't be picked up or retransmitted.

The radios are comprised of a tone key or an option to select the menu that lets you select a desired tone and also add it to the signal so that only the correct signals get picked up by the repeater.

To be able to achieve access over a repeater, you first need to tune your radio in a way that it is set to use the correct CTCSS tone and the correct offset. Offset is normally fixed for each band so that radios that have auto-repeater capability enabled can automatically have the right amount and direction of frequency shift. Regarding the CTCSS tone, choose your desired tone as applicable. By doing these two things, you are all set to determine if your signal can be heard by the repeater and if it can activate the transmitter. The process of activating the repeater is known as hitting the repeater.

First, cut off all the noise by squelch control adjustment. Next, always listen to the frequency to ensure you are not interfering with an existing conversation that's happening on that frequency. If the frequency is available, press the push-to-talk switch of your microphone to start the conversation and state your call sign, followed by "testing." Release the switch and check the strength of this short transmission on your radio. If your signal is successfully and properly received by the repeater, the indicator will show that repeater's output is received. This output signal is momentarily available, followed by a "tscht" sound—the sound that is due to the noise output of the repeater's receiver with no input signal.

Repeaters also "time out," which means sometimes a repeater suddenly shuts down in the middle of the conversation and produces no output signal. This doesn't mean the repeater transmitter has failed; this means the repeater has timed out and therefore isn't retransmitting the signals it received. If you wait for a while, you will notice that the repeater starts to respond again. But an important question here is—why does the repeater time out? Well, repeaters are equipped with a timer that begins when they start transmitting. And when the timer expires, the repeater turns off its transmitter. This solves two purposes—it allows the break-in stations to communicate by preventing one signal from occupying the repeater for long periods of time, and it prevents overheating of the transmitter. When the receiver doesn't see any input signal, the timer of the repeater resets itself.

To extend their listening range, and to be able to hear signals that are blocked by certain elements, repeaters use something known as remote receivers. These receivers operate so that the signal from them gets transmitted to the repeater's transmission region by an auxiliary station, for retransmission.

One repeater can be connected to other repeaters, which means the signal received by one repeater is shared by the other, retransmitting it over a wider area than what is generally covered by one repeater. At times, repeaters retransmit on other bands. This happens in cases where repeaters are located at the same time, ensuring they can be connected with the help of cables or control links. Control links are systems comprised of a transmitter and a receiver that sends only audio and control signals between the repeaters. The links carry signals that are capable of controlling various types of features, including enabling and disabling of signals by the linked repeaters and so on. The systems that comprise various linked repeaters (even if they are located far from each other) are known as repeater systems. How the system relays signals between these linked repeaters is something that can be configured using control signals.

If a ham uses a linked repeater system on a regular basis, he has an option to join the group that operates the repeaters. That way, he is authorized to use control codes to configure these repeaters along with performing basic test and maintenance operations, which is a valuable offering a ham can provide. The repeater ID is sent by a repeater controller, which also does the job of switching the transmitter on and off, and operates the timer. Today, we have microprocessor-enabled controllers comprised of sophisticated electronics, offering advanced features including weather conditions, voice announcements, etc. These advanced controllers are activated with the help of their own unique control codes. It is amazing to know that some of these repeaters also have the ability to make calls using ham radio; such a system is known as an autopatch. Again, these are some functions that are controlled with their own unique control codes.

To be able to use the control codes on the repeater, the transceiver should be able to generate the required tone or tone sequences.

If anybody looks at the repeater directory, they can see what repeaters are available and which are closed. If a repeater has a closed symbol, it is not available for public use and only authorized people can use it. Then there are repeaters that are dedicated for special purposes, including emergency communication. No matter what the case is—if it is dedicated for authorized people or for a special purpose, the owners prefer to have restricted use for the repeater, and this is allowed as per FCC rules and regulations. However, most repeaters are open and available for public use.

Then there are digital repeater systems. The internet offers a lot of benefits and hams have the advantage that they can roam freely; they are location independent and can operate from anywhere. So, the two, combined, can do a lot of great things together—and that's exactly what digital repeaters do.

Ragchewing Etiquettes

Now that you know a lot about ham radio, you know that it's not something that involves just spinning the dial and using a microphone. It requires a lot more than just that.

On the frequencies below 30 MHz, all bands have more or less the similar structure—Morse and data codes seem to spread across the lower third and voice modes are spread across the upper two-thirds of the band. The ragchewers can be found on every band; they are connected to the DXers on the lower part of the bands. Whenever any rare station gets on the air, several DXers make the ragchewers move up the band. The way they space themselves out around the data signals, the calling frequencies and the nets, helps them reap the benefits of the unique frequency agility of ham radio to find empty space in the band.

DXing happens at the lower ends of the band because of two reasons:

1. It is difficult to make long-distance contacts as compared to local or regional contacts. And since the signals get really weak over long distances, DXers find each other easily if they are operating from the same side of the band.
2. When the division of bands happened and sub-bands were created for different license classes, some high-class operators fell in the lower side bands. Distant stations took advantage of this non-uniform distribution of stations in the lower bands and started operating from there so that there was less of a crowd pursuing them when on air.

Since the way DXing works is different from other ham operating styles (DXing pulls more crowd), it is incompatible with other styles of nets (they are more orderly). And that's the reason all those styles gather on the other side of the band from where most of the DXing happens. With this structure, different group can operate without interfering with each other. As Morse and data code styles are incompatible with the DXing style, they continue to be in higher portions of the band. And since not many use Morse code and data these days, the data signals operate without interference.

One might think that the ragchewers are knocked about from various sides, but that's not really how it is. Ragchewing contacts happen all the time and that's why they occupy only the parts that are spared in the band. However, in case of disaster or emergencies, when a lot of nets are active, or when expeditions are taking place, or over the weekends when contests are running, the bands are almost full and it's difficult to find spare spots. Always remember that during a state of emergency, the FFC allocates certain frequencies in the band for emergency traffic and other operations. Ensuring these designated frequencies are always free is the responsibility of every ham. To let all the hams know of such declarations, the ARRL broadcasts the news and publishes it in bulletins, websites and so on, and the restrictions on the use of these frequencies remain in place until the FCC lifts the state of emergency.

Ragchewing is generally found in the repeater section of the UHF and VHF bands, even though there are many open spaces available for conversation in the weak signal portions of the bands. Below, you'll find a list of frequencies and their respective portions of VHF and UHF bands.

Band	Frequencies assigned	Purpose
6 meters	50.0—50.3	CW and SSB
	50.070, 50.090	CW calling frequencies
	50.125 and 50.200	SSB calling frequency, use upper sideband (USB)
2 meters	144.0—144.3	CW and SSB
	144.050	CW calling frequencies
	144.200	SSB calling frequency, use upper sideband (USB)
222 MHz	222.0—222.15	CW and SSB
	222.100	CW and SSB calling frequency, use USB
440 MHz	432.07—433.0	CW and SSB
	432.100	CW and SSB calling frequency, use Upper Sideband (USB)

The lower parts of the UHF and VHF bands are known as weak signal portions. The reason for that is because the contacts that utilize the CW and SSB can be established with weaker signals as compared to FM. Most of the weaker SSB and CW signals that we get to hear are quite strong for making great contacts.

Another important thing is knowing when to chew. Like everything else, ragchewing has its good days and bad days, be it HF or UHF/VHF. When you want to connect to any station, you can let it know that you are looking for an extension in a number of ways. Once again, the best way is to listen. If you want to know how to extend a call successfully, listen to learn how others do it.

When you are tuning on an open band, look for the social aspect of your contact. Normally, weekdays are considered to be good for ragchewing, particularly the daytime, when amateurs are at work. However, the timing might be different for hams, depending on which part of the world they are located in. You can always look for the details around what band is the best to use at a given time, although a repeater or the lower band can always be used if the contact has to be made within the same region. To make coast-to- coast contact, the higher bands are preferred and the better the antenna, the more options you would have. Even over weekends, lots of hams operate. So don't be shocked if you see a full band on Saturdays and Sundays. However, the fact is that there are many hams on air during this time for you to contact. If you learn about a band that is fully dedicated to an event, you can always find a quieter place in another band. For instance, the WARC bands are always available for casual operating and ragchewing since they are never used for any contests.

Now, the bad times. Ragchewing happens over a long time, so one can select a time and band that offers stable conditions. And knowing the fact that propagation changes with the position of sun, one has to take into consideration various parameters for deciding the appropriate time. For instance—noon is a bad time on the higher bands. So select the time based on your experience. Again, knowing that events and other activities take place over the weekends, always consider the calendar to see how occupied the bands are. This will save you from surprises when you are on air.

During the not-so-good times, here are some of the tips and strategies that can save you:

- Most contests and nets take place in the traditional bands. For instance, the WARC bands are mostly empty for a QSO as the events never happen on the band.
- Some contests take place on both CW as well as over the phone. There are hardly any events that are conducted on all the bands—CW, data and phone. This gives you an option to switch between the modes to find the available band.
- You must know how to operate your receiver properly. You must know how to operate and use controls such as passband tuning, IF shift and so on. If you know how to use your filter settings, you can efficiently remove the noise and interference by simply making the right adjustments in the receiver.
- Always be ready with a backup plan. There is no assurance that you will always find an empty spot in one of the bands on a given day. That's why it is always important to have a backup plan ready, in case you do not find a clear frequency. Hams are given frequency freedom like no one else, so do not hesitate to use the big knob right in front of you.

Okay, so you found a frequency to ragchew and you are tuning into that band. But what if another station wants to ragchew? In such cases, the easiest way to identify if another station is interested in ragchewing is breaking in. If not, you can wait for the station to finish.

If you do not want to go with this option, another option available is CQ. Based on voice, you can find cues around another station looking to ragchew. If you hear an easy tempo of speech and relaxed tone, it is an indication that you should look for another station. So consider a station that has strong signal—not too strong, but something that has steady strength and is easy to copy. The best ones are those that last for enough time so that a signal holds up. One of the cues that a station is not interested in ragchewing is a targeted call. Another tip is a call that has numerous stations responding. This station might be being used for a contest or an event. So continue, if you wish to ragchew.

Round tables are another great way of ragchewing. This happens when more than two hams contact each other on a single frequency, just the way it happens in round-table conferences in which each person takes turns sharing their point. In similar fashion, stations can join and then sign off, and join back in whenever they wish.

As you tune across the bands, evaluate various techniques and see what you like and don't like. Go for something you like and improvise it, as that is how the amateurs work.

Chapter 7—Build a Home for your Ham Radio

If you are running a ham radio station, you will have to spend a lot of time in your shack/station. It does not matter if you are an amateur or a passionate and experienced enthusiast. It is imperative that both your gear and you make yourselves as comfortable as possible in your quarters. Towards this end, you should be prepared to be dissatisfied with the initial layout! For instance, do you find very little room for maneuvering? Are there stacks of equipment everywhere? Poor lightning, faulty antenna setups, etc. may hamper you. Whatever the case, refuse to make adjustments and remain satisfied with the way things are. Only then will you take serious action to experiment with everything and keep improving your style of operating! In fact, you should ensure that your manner of operating and preferences always remain highly efficient and comfortable! Towards this end, we'll share a few tips, which should come in handy at all times.

Getting the Ergonomics Right

Ergonomics is a science. It refers to the creation of workspaces and equipment such that they display maximum efficaciousness and effectiveness. This science also takes into consideration any operator/worker's fatigue and discomfort. This is extremely important, since you are going to spend hours sitting in front of your ham radio. You should feel comfortable, but you can't do so if you have to twist your body into awkward positions due to the furniture not being of a suitable height. Similarly, harsh lighting can hurt your eyes. These are but a few examples, but many things can cause physical discomfort. Therefore, it is imperative that you get all your ergonomics right.

At the outset, figure out your main objectives for setting up the station. Ask yourself why you want to succeed at your new hobby/work. You may love to concentrate on digital modes, HF ragchewing on the phone or monitoring VHF FM repeaters. Whatever the case, every focal point merits a different approach. Then again, you are not going to stock your station with diverse types of equipment. You will obviously opt for mostly adhering to a sole piece of equipment. You may use this equipment to decide how you would like to arrange your radio station. Would you like the focus to be on the computer keyboard and monitor, Morse code paddle, the radio or the microphone? If you gain clarity in this area, you should find it quite comfortable to operate your ham radio station. To illustrate, if you are someone who loves to function with a limited number of monitors, transceivers, transmitters and receivers, it helps to have a U-shaped layout. This helps your hands be able to reach everywhere without much difficulty. Of course, you would receive additional help if you were to place the most frequently used items (knobs, switches, etc.) within easy reach. Similarly, do not stack your equipment beyond 18 inches in height, for you will then have raise yourself on your toes to perform operations. If you do not like a U-shape, you may opt for a corner layout. However, this will do only if you have a comfortable corner available in your shack. A third option is to create diverse workstations for diverse types of equipment.

Whatever you do, take care to keep your posture correct and take much-needed breaks at the right times.

The Operator's Chair

When you are ready to spend money on the best of equipment and accessories, why skimp on spending for a comfortable chair? Do not reward your hard work with a wobbly and cheap specimen from a garage sale. Note that if you cannot position yourself properly during the long hours that you spend at your shack, you will not prove to be a good ham radio operator. Your mind will eternally concentrate on making yourself more comfortable!

If you really want a great-looking and comfortable chair, go for a roll-around type of specimen, which offices commonly use. There should be healthy support for your lower back, as well as solid padding on the seat. If there are levers to help in raising or lowering the height of the chair, it's even better! However, ensure that there are no arms on the chair. You may have to lean over frequently while conducting operations. Your lower back will face unnecessary stress if you have to stretch your torso over the chair's arms all the time. That said, you may not find chairs without arms so easily. You will therefore have to remove them after purchasing the chair. Instead, look for a top-of-the-line secretarial model. This kind of specimen permits easy rolling and swiveling. True, the cost may go a little higher in comparison to what you originally planned to spend. However, your back will thank you for keeping it comfortable! Search the Yellow Pages for used furniture. You should be able to obtain good bargains.

You may decide to use a footrest, if you soon get tired of resting your legs on the floor. However, it must allow you to control your switches and knobs comfortably, without having to lean over or stretch too much.

The Operating Desk and Shelves

Apart from using your chair frequently, you will also befriend your operating desk throughout the day. In fact, the surface of the desk is more important to you than the rest of it! However, it is your choice of how you wish it to appear—well organized and roomy, or stuffy and stingy! If you want to be neat, you will prefer a desk that offers sufficient space for everything that you want to place on it, in an attractive manner. It is good to opt for a specimen that is around five to seven feet in length and at least 2.5-3 feet in depth. With this, you should have sufficient space to operate your small electronic items. Alternatively, you may have a habit of cramming everything into the smallest of spaces possible. This could be because you hate spending money on good quality and sizeable desks, or because you feel that having all your gear spread out does not look nice. Note that too much cramming and too little space can result in claustrophobia later on!

It helps that diverse types of desks, with varying heights and depths, are available in the marketplace. However, you must first figure out if your ham radio will go on the desk directly or onto a shelf just above it. Only after deciding this will you able to decide what kind of a desk you want. For instance, if you love having a keyboard on your desk, you must be able to rest your forearms comfortably on the desk too. Your upper and lower torsos must feel comfortable. Think about it! If you want to fiddle with the radio a lot, such as operators in high-frequency stations do, you cannot afford to have your arms resting on the edge of the desk or on your elbows all the time. You will find yourself crabbing about pain all the time! This is because the keyboard is occupying so much space. Therefore, you will need sufficient space for supporting your wrists while operating your keyboard, and will need to hunt for a desk that will prove suitable.

Generally, a desk is structured to be at a height of anywhere between 28 and 30 inches from the floor. With regard to depth, the minimal requirement is 30 inches, in case the transceiver is the typical type needing frequent tuning or adjustment, and sits directly on the desk. There should be a distance of at least a dozen inches between the tuning knob on the transceiver and the front edge of your desk. You will have to rest your entire hand on the desktop in order to reach the controls. In case you plan to place your radio on a shelf located just above the desk, ensure that your hands are able to reach it comfortably without having to overreach. You should be able to view the controls properly, even if they are at a distance.

Bearing all these considerations in mind, you might just opt for a computer workstation and be done with it. It is admirably suitable for a shack that is small or moderate in size. True, you might have to add a few shelves to this workstation, but you should be able to utilize the major framework wonderfully well. The only snag is that your arms may not find the depth suitable, specifically if you use them a lot for your style of operating the ham radio.

In case you still want a small desk or table, there are ways to make it look larger. You do not even need to spend too much money on the expansion process. For instance, you may create an addition from good quality plywood and attach it to your table or desk. It is possible to enhance the length and width of your table/desk with the aid of Masonite sheets, trim moldings, corrugated nails, etc. All that you require is to awaken your latent creativity!

With regard to shelves, you may not need too many of them during the initial stages. In fact, even a flat table surface should suffice for some time, just to keep your amateur station going. If needed, a couple of shelves may come in handy. However, once you complete a few years of ham radio operations, you will need to give serious thought to the idea of shelving. Otherwise, where will you stack all your paperwork? There will be logs, books, manuals and so on. Furthermore, you may need to store more radios, accessories, spare parts, etc. Towards this end, therefore, consider using pinewood to construct some shelves. They may even be part of your table arrangement, directly connected to it. In fact, if you need a good amount of storage space, ensure that you buy a couple or more of modular units, all possessing strong, supportive legs.

Basic Equipment

True, you may be keen to launch a ham radio station because it is your hobby. Then again, you may want to move from amateur to expert by learning everything you can about ham radio operations. However, there is no denying that you are keen to make contacts or QSOs. Towards this end, some basic equipment comes into play for making your contacting efforts a complete success.

Transceiver

It combines both a transmitter and a receiver in its structure and functioning. In fact, it is a better choice in comparison to obtaining a separate receiver and transmitter. As an amateur, you cannot afford to learn how to operate both successfully. The cost of the transceiver is dependent upon the output of power and the frequency range. Naturally, a good quality one will be more expensive, especially if it is from a branded establishment, such as Yaesu, Kenwood or Icom. In case you are planning to go in for VHF operations, wherein you prefer to converse with people using repeaters, then you will be able to make do with a small mobile handheld receiver. Just ensure that it possesses all the necessary components.

Power Supply

This grants power to the transceiver. There are diverse types of power suppliers. Research what is most compatible with your type of transceiver. For instance, if your power supply does not offer sufficient voltage, your transceiver will receive too little power to function properly or even turn off. Alternatively, if you purchase something that gives out too much of power, the power supplier could blow a fuse. Alternatively, it may damage your radio. In case you are still confused about your power supply options, just ask an experienced ham operator to take over.

Antenna

This item is affordable, in comparison to other pieces of equipment. You may obtain a vertical antenna, a G5RV, etc. It is imperative that you check with your local area's zoning laws prior to setting up your antennas. This is because some towns have certain restrictions in place regarding the type of antennas you may put up.

Antenna Tuner

Your tuner should be able to match your antenna to the transceiver. If they do not appear compatible, you may end up blowing a fuse or causing serious damage to your transceiver. This is because you are not able to control the power that is coming into play during transmission on your transceiver. You may opt for a manual tuner or an auto tuner. The quality of each is dependent upon the brand, as well as other factors. Since you are still in the process of acquiring expertise in ham radio operations, it would be best to purchase a manual antenna tuner.

Key or Microphone

This has a link to the mode of communication that will come into play whenever you begin your radio operations. You will require a key to communicate via Morse code. You will require a microphone in order to send SSB or voice. While several transceivers come equipped with diverse microphone jacks, there are also many transceivers which possess compatible microphones. Keys related to transmission of Morse code, however, do not offer the same guarantee. You will have to choose between different types of keys, such as bug, straight, etc. It would be best to choose straight keys, since they will suit your 'simple' expertise with the ham radio.

Extra Cables

They need to be extra-long since you will have to connect quite a few pieces of equipment together. Towards this end, each may be one to five feet in length.

Purchasing New/Used Equipment

You may be fond of purchasing new equipment. After all, it promises to function without hiccups, and it smells good too! In case it does not live up to its promises, you may contact customer service representatives and request their help. They will render assistance because your gear comes with a service warranty. Similarly, sales personnel, who are happy to receive new customers, are more than willing to provide guidance regarding the setting up of your radio, its accessories, etc. In fact, they may even share application guides or technical bulletins with you, such that you do not have to worry about performing popular activities at your ham radio station.

With regard to the locations of stores that supply new equipment, you may peruse various journals and publications focusing on ham radio operations and stations. Many dealers place advertisements every month in magazines, such as *World Radio, QST*, etc. Sometimes you may come across virtual catalogs on the internet. Some websites prefer to offer lists of smaller shops, as well as local stores.

While it is definitely safer to depend upon brand-new things, there is no denying that used equipment may also prove to be an excellent bargain at times. The only snag is that you must have the experience and expertise to be able to recognize this bargain! Regardless, as a ham radio operator, who is trying to create a foothold in an entirely new world, don't resist if something does come your way! You will be able to save some money. Therefore, peruse certain swap sites online, such as eBay. You should be able to find other ham radio operators offering equipment for sale. Then again, you may go through the classified pages on other well-known web portals. Thus far, you may have tackled everything on your own. However, when it comes time to actually purchase such things, please request the assistance of an experienced ham radio operator.

There is another way of purchasing the necessary goods for your radio station. This is mail order. However, unlike purchasing online, you will have to shell out some money for shipping. In case your equipment requires servicing, you will have to pay for the shipping once again. Then again, there is no guarantee that the seller will have all the spare parts or replacement parts which your equipment requires. Thus, you end up spending so much on your mail orders that you feel it would have been better to purchase everything from the local radio store! Regardless, do not discard mail order services as if they were useless. Instead, opt for a healthy balance of mail ordering and local shopping, such that your ham radio station feels satisfied. Otherwise, if you depend purely on mail orders, you might find yourself stuck in the middle of construction or setting up a project and find you have to keep waiting for the materials to arrive.

Placement of Gear

How will you place your equipment? Will you just dump everything wherever you see a small space available? Of course not! You are well aware that convenience is a keyword in operating your ham radio station efficaciously. Therefore, you will go all out to ensure that your gear, which will require adjusting at all times of the day and night, is always within arm's reach. These units include transceivers, keyer, transmitters, transmatch and receivers. It makes no sense to get up from your chair every time you want to adjust something! You should place the transceiver or receiver in such a way that your left or right hand is able to tune the knob easily. You should be able to adjust frequency easily and accurately. However, this is possible only when your eyes are clearly able to view the digital frequency readout displays or analog. Whatever accessories you do not require so frequently, you may keep to the left or right of your major focal point.

Similarly, you need to award specific attention to your hand keys, paddles and bug keys. Towards this end, ensure that your table/desk has sufficient space to support your arm easily, right up to the elbow. Your fingers should be able to touch the keys comfortably, without straining the 'sending' arm, which will be busy handling continuous waves (CW) while transmitting or communicating. In fact, your fingers should be the only 'tools' to come into play for manipulating the keys. When you find a particular spot that affords complete ease for sending, place the key there. Fix the key to the spot with the aid of pads or adhesive-backed tape. Sometimes you acquire a 'glass' arm due to over-sending! You may alleviate this problem by elevating the paddle/key. Place it on a higher platform, such that is slightly above the desk/table's surface.

In case, you are content to make do with just one radio, there may just a ground wire, a feed line and a power cord in place. However, as you collect more radios, a personal computer with monitor, tuners, amplifiers, and all manner of electronic devices, you will see a conglomeration of wires trailing all around the shack. Even your arrangements will contribute to the 'wire' mess! Therefore, the first thing you should do is to provide labels to every single wire, such that you are able to differentiate amongst them. Additionally, as your interests and focus change over the years, you will be able to move around items easily, without becoming confused. At the same time, you must bunch up the wires and tie them up neatly. Ensure that they never get in the way of your feet by pushing the bunch into a corner or any other safe place.

The Computer

It is unthinkable to be without a computer in the twenty-first century, regardless of whatever profession or business venture you take up! It is no different with a ham radio station. Surprisingly, many ham radio operators have embraced the advent of a computer into their respective shacks quite ardently! Earlier, operators considered computers useful replacements for paper logbooks. However, they have gradually taken over other activities over the years. Today, the computer is akin to a second operator, which controls the radio, sending and receiving continuous waves, linking to thousands across the globe via the internet, etc.

The majority of computers that enter hammers' shacks function on Windows-based machinery. The vast array of software prefers to befriend Win9x, Windows 2000 and XP operating systems, since they prove most suitable for ham applications. Similarly, quite a large number of MS-DOS applications, which are compatible with MSDOS windows, favor Win9x and Windows 2000. However, digital mode enthusiasts are going for Linux in a big way. Even Macintosh is foraying into this field. This helps ham operators to access various kinds of programs for performing routine and exclusive tasks. You should be able to access relevant websites over the internet, which provide detailed information about everything.

To sum up, it does not matter what you prefer with regard to the platform or the operating system. You have the good fortune of being able to access software programs and tools for every kind of operation. Commercial businesses supply some of the software. However, keep in touch with the amateur ham radio community as well. This community has come up with an astounding quantity of freeware and shareware. Many ham radio operators do not hesitate to share their expertise in diverse ways, or even to develop software for performing popular ham activities.

Since the focus is so much on computers, it might help you to know what kind of a machine you should have in your shack. To begin with, you would do well not to purchase an old or leftover computer. Note that the world of computers and the internet is advancing rapidly. If you decide to skimp on an important piece of equipment, you will end up with an underpowered computer possessing an obsolete version of some operating system. Set aside a part of your budget for a brand-new computer with a Windows operating system. You will be able to obtain all manner of advanced software for your ham operations. Here are some recommendations which might come in handy.

- **Expense** – It is much cheaper to buy a configured/assembled computer, rather than something readymade. This should definitely help in cutting down costs.
- **Desktop computer** – Your radio station would much prefer a desktop computer, rather than a notebook. You can even place it on the floor of

your shack and not worry about it too much! It costs much less than a notebook computer. As for the monitor, obtain a model that will not harm your vision. A 20-inch monitor should do.

- **Notebook computer** – Yes, you can opt for a notebook computer. However, just ask yourself if you will be comfortable staring at a small screen throughout the day. Then again, where will you place it? In case you plan to share it with other computing needs, you are going to have a hard time connecting and disconnecting it repeatedly. Various radio interface devices have different connections. Regardless, if you still insist on a notebook, ensure that an external monitor, mouse and keyboard for easy operations accompany it.
- **Storage and performance** – It helps that the software programs linked to your ham radio operations can make do with less computing power. A powerful computer should suffice to help you run several applications at the same time. You do not need a high-end machine. As for storage, you should find a standard rotating disk to be adequate for your station. An SSD system drive is great for speedy operations.

Placement of Computer

Despite setting up a radio station, it is inevitable that you will turn to your computer more than you do to your radio! In fact, the computer monitor is an extremely prominent piece of equipment resting on your desk. Furthermore, you should position it for quick operations. Therefore, you need to follow certain guidelines to utilize your computer in comfort. Towards this end, it is necessary that you place the desktop at the appropriate height, such that you can take up prolonged typing, even for extended durations, without tiring your arms or back. An alternative is to buy a keyboard tray.

When you purchase your monitor, ensure that it is of high quality. Additionally, have it placed at a suitable distance and height such that you can focus on your monitor in a relaxed manner.

To illustrate, if your monitor is too far away from your eyes, they will experience strain. On the other hand, if it moves too far to the right or left, your back will never forgive you! It is important that the distance between your eyes and the monitor remains anywhere between 18 and 24 inches. However, these dimensions are undergoing modification in modern times. Today, people believe that the greater the distance between the monitor and your eyes, the less the strain that your eyes incur. Furthermore, you will not experience neck pain.

Another recommendation relates to the height of the monitor. The monitor should rest on the desk/tabletop in such a way that your eyes look down upon it. In actuality, the screen should be set at an angle such that it does not confront glare from natural light, artificial light, or a task light. It is best if the monitor screen appears 10 times more brilliant than the light level surrounding it. If this is not possible, opt for eyeglasses which are task- specific in nature. Spend some time getting your computer ergonomics right, such that your long hours at your operating desk prove enjoyable.

Now, how will you place your keyboard? The majority of people place it directly in front of them, on the desk/table. However, if you prefer to write more than you type, you may place it at an angle on one side. Leave the mouse beside the keyboard. You will be able to find it easily. In case you do not like such arrangements, you may opt to place your keyboard on a slide in and slide out tray. The tray is underneath the desk/table, such that it remains at the correct height while in use. Generally, the tray tends to be at least 36 inches in width, enabling it to hold a pencil tray with various items and the mouse on it, too. It is imperative that your wrists remain straight while you type. Therefore, the keyboard should afford your wrists some space.

Lighting

Do not skimp on this important aspect of your ham radio station! How will you be able to read meters and dials accurately if your shack looks like the lounge of a cocktail bar? Then again, you need to be able to see what you are writing when you log in or are keen on taking specific notes. Note that whenever you read or write, you will require lighting that displays five times more illumination than that which general lighting levels display.

If you relax in a dim atmosphere, you are apt to feel sleepy and sluggish. Bright lights will ensure that you stay alert and awake. Whatever lighting you use, ensure that it reaches your desk over your left shoulder. You may even add extra ambience to the brilliant atmosphere with a desk lamp. The task-specific lighting should not produce a glare and needs to be diffuse (not produce shadows).

Furthermore, there should be a healthy balance of colors. There are all kinds of lights today, each offering unique features. Consult an expert for good advice.

Other Equipment

Feed lines

They have another name too, which is transmission lines. These are the conduits which help maintain the radio frequency power between the radio and the antenna. In other words, generated energy travels from radio to antenna via a feed line. The majority of ham radio operators feel inclined to buy a coaxial cable, which is generally known as a coax. The couple of circular conductors, which reside on the same axis, contribute to the origin of the name. In fact, one conductor lies inside the other. The inner one is the center conductor. Surrounding it is the shield, a multi-stranded outer conductor. In turn, an insulating plastic jacket encompasses the shield. Insulating material is present between the two conductors too. It may be just space left for air, or an insertion of foam/hard plastic.

The coax feed line may be 52 or 75 ohms, RG-213, etc. The reason for its popularity is its flexibility. You are welcome to experiment with however you wish to feed your antenna via this feed line from your shack! It is your choice to run the feed line through the window or wall of the shack to the outside. If possible, you may use your attic too. However, just ensure that the feed line does not radiate into other electronic devices stationed nearby. You should also use a coaxial cable safely in places where children and pets love to hover nearby. You need never worry about them while undertaking your transmission duties.

While all this sounds good, there is a drawback too. Sometimes, when you may be operating at frequencies going beyond 30 MHz, if your feed line is over 100 feet, you may face some loss of power. In such a scenario, you may opt for open-wire feed lines. They include 450/600 ohms ladder lines, TV twin leads, etc. Ladder lines refer to two parallel wires with some insulating material separating them. Operators who go in for higher frequencies prefer them for diverse antenna designs and random-length dipoles. However, these feed lines are tremendously susceptible to metal objects. Therefore, you will have to ensure that the feed lines run through your attic. In case you still decide to opt for your shack's wall or window, make sure the exits are generally out of the way. Since the feed lines carry a lot of voltage, it would be best to keep pets and children away from them as much as possible.

Maintaining a Notebook

Use a spiral-bound notebook. Alternatively, opt for a three-ring binder. You do not need anything elaborate to maintain records. Strong winds will not affect bound notebooks. As you get into the process of building your shack, enter every detail into your notebook. Make sketches of every design that you think of, and list out the equipment and accessories required for operating your ham radio station. To illustrate, you might wish to connect two pieces of gear with a cable comprised of multi-conductors. What is the color of each wire, and where is the attachment? Please note down the tiniest of details, for they are important for efficient operations. Striving to remember everything later may prove terribly tedious and frustrating.

Also note that your 'recording' job does not come to a halt as soon as the station is up and operating. You are bound to keep adding new gear and setting up novel connections. Your antennas will work differently at different frequencies. Even the wiring patterns linked to the tiniest of gadgets are important. In simple language, your notebook should be a valuable mine of information! It should be akin to an encyclopedia that will help you handle everything at your ham radio station wonderfully well!

The Log Book

It is actually a complement to the shack notes that you maintain at your radio station. You may use a binder or notebook (a small black one should do fine), wherein you make manual entries for every contact that comes your way. Whenever you need to troubleshoot or evaluate your gear, you should find it handy to refer to your logbook. Do not forget to update this notebook regularly.

What exactly should go into this record?

As a ham radio operator, you may have a tendency to use World Time (Coordinated Universal Time or UTC), forgetting about everything else. You need to keep time for your local contacts, as well as your radiogram identification.

Note down the frequency by keeping track of the band. You may use wavelength or MHz as units of measurement.

The call sign refers to a record of every radio station that you connect with during your duty hours.

Thus, time, frequency and call sign suffice to let people know about the who, when and where of your radio station. Apart from these basic recordings, you might wish to track other things too. These include the mode that comes into play, requisite personal information about another ham operator and a report about signals sent and received. Note that you need not limit your conversation to mere exchange of records maintained in your logbook. You are welcome to exchange interesting tidbits of conversation too, which have little or nothing to do with a radio station!

Here is another suggestion. If you are highly active at your radio station, you might find it easier to maintain a log on your personal computer. It makes it easier to look up earlier contacts that you have created with another operator or at a radio station. Your logging program should prove highly useful for creating a day-to-day radio diary. This diary could be a web blog, which enables you to keep track of local weather conditions, regular behavior and performance of your gear and ionospheric and solar conditions. You will be able to use these programs to export data via a standard file, such as automated data interchange format (ADIF). With regard to portable or mobile operations, you may be able to use portable logging programs. For instance, something like GOLog functions on Palm OS PDA. Use this website, home.earthlink.net/%7Egolog/ to connect to the program.

The Radio

There are all manner of radios of diverse sizes and shapes, with accessories that are just as wide-ranging. Therefore, you may not adhere to any hard and fast rules for setting it up or operating it. For instance, if you are an HF operator, you will be engaged in plenty of tuning operations. Naturally, the placement of your radio will have to be perfect. In contrast, if you are a VHF-FM operator, you will not engage in too much tuning. Therefore, you may place the radio at a distance. In fact, you can just place the equipment to one side of your computer keyboard or monitor. This will make operating easy for you. In case you are right-handed, you can place the radio to your left, with the trackball/mouse remaining on the right. In case of the reverse, the mouse/trackball will go on the left, and your radio will be on your right.

A large radio possesses a bail. This is a folding wire support. Sometimes a radio possesses adjustable feet. The bail is capable of folding flat against the underside of the rig during transportation. If you adjust the feet or the bail, you should be able to raise the front portion of the radio. This way, you will be able to place it at a good angle for comfortable viewing. Without even moving your head down or up, you will be able to see the labels on all the controls.

You may have to do some trial and error experimentation in order to discover what works best for your ham shack. These are but a few samples of how you could arrange things in your shack. Just stretch your creativity a little, and make the place seem like a second home!

Taking Charge of Radio Frequencies and Electrical Safety

It is imperative to bear the basic principles of safety in mind, regardless of whatever kind of radio station you put together. Plenty of literature is available in this arena to give you any help you need. Do not take this aspect lightly, for your equipment might incur damage, or someone might be seriously hurt. Outlined below are the safety fundamentals that you should take into consideration. There is nothing complicated about the whole thing. It is just a matter of following some simple rules.

Rules and Regulations

To begin with, acquaint yourself with the tables and rules that the National Electrical code, or NEC, recommends about dealing with the wiring for obtaining AC power. Apart from the NEC, you should also be able to find plenty of training or how-to guides at your local library, or even at centers that cater to home improvement needs. Regardless, if you still don't have confidence in your ability to do a good job, ask an experienced electrician to take over.

Handling DC

You may not know how to tackle DC power, specifically if you are using a car. There is always the possibility of poor connections causing damage, as well as short circuits occurring suddenly. You cannot take such situations lightly, for they lead to the spread of fire. Then again, even your ham radio will function erratically. Therefore, peruse the safety literature thoroughly, and tackle the jobs on your own. Alternatively, hire a professional, who can take care of the installations properly.

Family Safety

Your family needs to believe that it is safe while you go about setting up your radio station. Towards this end, ensure that no electrical circuits remain exposed to any accidental touches. Whenever you plan to work on electrical equipment or wiring, make sure that a safety lockout is in place on every circuit breaker. This will prevent the circuit breaker from closing and energizing the circuit. It also helps to have functional fire extinguishers around. Instruct the members of your family regarding the safe movement of power from the shack housing the ham radio.

Lightning Strikes

Is the area of your residence prone to lightning strikes? Note that lightning has immense potential to destroy and cause rapid fires. Lightning is akin to a wave pulse possessing extremely high energy. Consider voltage fluctuations. These fluctuations possess high-frequency energy since they are very rapid in nature. When this energy reaches peak levels, it involves dozens, or even hundreds, of kilohertz. The major part of the energy moves anywhere between low alternate current frequencies to something as high as 1 MHz. If such strong energy spreads to hundreds of megahertz, it is bound to cause immense damage.

Now, put lightning in place of voltage fluctuations. It is similar to an energy source possessing very high-frequency energy. However, the bulk of its energy remains in the range of lower frequencies. Whenever there is a lightning strike, it is as if a million volts of electricity have struck the sky! In reality, thousands of amperes of current are flowing through lightning. Therefore, it is imperative that the ground be ready to take the brunt of it. A healthy ground should possess extremely low impedance, which spreads over an extremely wide frequency range. In other words, the pathway for everything conductive (having the power to conduct/transmit) which may enter the station/shack must be one of common low impedance.

All the television antennas, metallic conduits, power lines, pipes and telephone lines on display should share the common ground connection bus, with its extremely low impedance. Naturally, thin wires will not work in this kind of a scenario. You may not opt for braided conductors, which are loosely woven. However, you may require it if the area where your radio station is located requires conductor flexibility. Therefore, the best thing is to opt for wide and smooth-surfaced copper flashing. Alternatively, you may choose heavy and round copper. This is because the RF resistance per unit length of a selected surface area is low in round copper. In contrast, wide and flat copper exhibits less reactance, as well as lower overall impedance. It further helps if you possess a marvelously good ground, which exhibits a wide, solid and smooth surface.

Heights attract this source of immense electric power from the skies! That is why power lines become favorite targets of lightning! Power lines connect to utility lines. Therefore, lightning follows the same path, reaching the house. It then travels the length of the house wiring, touching all sorts of equipment along the way. Finally, it moves to the ground via the antenna systems. Regardless, the house wiring and equipment cannot remain free of damage or danger. It is the same with tall towers, surrounded by significantly high structures. That is why experts stress so providing grounds with low impedance pathways in order to prevent lightning disasters from occurring.

It is imperative that you take all possible steps to protect both your home and your radio station. For instance, you might opt for disconnecting all the feed lines of your antennas whenever they are not in use. Then again, you may choose to implement professional grounding or bonding. Note that many experienced radio operators opt for at least a lone lightning arrestor in position. They use this option, regardless of how they feed their antennas, in order to prevent lightning from ever reaching your shack. If you would like to emulate their example, ensure that you obtain something that is fast-acting and suitable to handle the maximum power generated by your radio operations. This could be a gas tube fuse or GasFET possessing connectors of the PL-259/UHF or Type-N kind. You could also have another lightning arrestor placed near the exit of your feed line from the shack.

It should help you to know that lightning keeps a keen lookout for a pathway that offers the least resistance. Therefore, have a coax line buried in the ground. When lightning strikes, the coax cable will help it to break its skin and travel into the ground. The cable will not permit lightning to travel along its length and enter your shack. True, you can never predict where lightning will strike, or even guarantee your shack's protection 100 percent. However, there is no harm in taking as many steps as possible in order to prevent disaster. Just make sure that whatever steps you decide to take, they reach completion in an appropriate and safe manner.

The Human Body and RF

There is no guarantee that your transmitter will always generate safe signals. They may prove as hazardous as the direct shocks that occur with contacting live circuits. Surprisingly, the human body is capable of absorbing the energy emitting from radio frequencies (RF). In turn, this energy converts to heat. Of course, the energy generated varies in alignment with the kind of frequency coming into play. For instance, amateur signals cannot cause any real harm to the human body. However, if the antennas become responsible for focusing the signals in such a way that RF hits your body for a long time, you will face harm. The maximum hazards occur in the UHF and VHF regions.

To illustrate, a microwave oven is capable of generating maximum UHF frequencies. You have to handle them with caution. As mentioned earlier, study useful literature, genuine websites and consult expert professionals to ensure that you take all the necessary precautions for operating your radio station safely. Also, while setting up your ham radio station, keep the radio frequencies' safety guidelines in hand. This will help you evaluate safety, step by step.

First-Aid Kit

It is imperative that you have certain first-aid skills at your fingertips, as well as a first-aid kit handy. Even your family members must know how to utilize the first-aid kit, whether you have placed it at home or in your shack. It would be good for all of you to have training in cardiopulmonary resuscitation (CPR) too. You never know when someone's heart will start causing problems and you will have to take emergency action.

Grounding Power

It is imperative that there be healthy grounding for all the radio signals that come your way, as well as for both DC and AC power. However, it is easier to tackle the former in a straightforward manner, rather than the latter. Nonetheless, you will have to tackle both.

The term 'ground' has nothing to do with the earth. It just refers to the lowest common voltage that may affect electrical equipment, thereby affecting their functioning. To illustrate, all power systems remain at zero voltage on the earth. However, voltage may fluctuate due to various reasons. Therefore, grounding ensures that every piece of equipment retains the same voltage reference. Thanks to the grounding process, it is possible to keep DC and AC wiring safe. It becomes possible to connect exposed conductors, if any. Such exposure is visible in equipment cases. Here, the experts connect everything to a zero-voltage point, or directly to the earth. The zero-voltage point may be a building frame. Whenever you opt to guide the current away from you during a short circuit occurring between the exposed conductors and source of electric power, you and your equipment remain safe. The low voltage ensures that there is no electric flow between diverse pieces of equipment, even if they are touching one another directly. You will not be affected either, should you touch both pieces of equipment at the same time.

The 'third' wire is the dedicated conductor which brings about grounding for power safety. At your home, you should be able to note AC outlets with a three-wire pattern, which serve as connections between the ground pin and your home's ground. This connection is visible at the master circuit-breaker box. As AC power experiences very low frequency, you need not worry about the length of the ground wire. However, make sure that it is sufficiently heavy to tackle fault currents, if any.

As for DC power, the voltage goes as low as 30 volts or less. In this case, it is imperative to reduce the flow of excess current, as well as poor connectivity in electric circuits. They generate extreme heat, thereby increasing the chances of starting a fire. Preventing shocks isn't very important here. Instead, you must pay close attention to the sizes of the conductors, as well as keeping connections clean and tight. Similar to the AC power system, it is not necessary to worry about the length of the conductor. Just take note of its capacity to carry current.

Grounding RF Signals

Sometimes, the power-safety techniques that you utilize for DC and AC systems do not work as well with high frequencies at ham radio stations. With regard to a radio frequency, the ground wire need not be lengthy. You have to ensure that it behaves like a transmission line or antenna. For instance, if you use a wire measuring eight feet for 28 MHz, it will provide a wavelength of 1/4. As a result, the voltage may prove to be very high at one end, while being very low at the other! This is definitely not something you want for your radio station!

Obviously, it is not easy to have one common point providing low voltage at a radio station. The wiring system will not be able to cater to every piece of equipment at your station. Therefore, you will have to keep all the equipment functioning at the same voltage, similar to what all other station builders do. Towards this end, you may use a single strap or heavy wire to place behind the equipment or underneath it. This is your RF (radio frequency) ground. Next, connect every piece of gear to this heavy wire via short strips of copper braids or straps. In fact, you will discover that your ham gear possesses ground terminals which are perfect for this purpose.

You should be able to find copper straps at roofing stores or hardware shops. The dealers refer to them as flashes. Do not spend too much on them by wandering into expensive stores. Just find a surplus metals dealer and poke around his/her shop until you find what you want! As for copper sheets or strips, you should be able to find them in diverse sizes and gauges. If you are lucky, you may even discover pre-drilled, evenly-spaced holes in a copper bar! This will be akin to a ground bus! Whatever the case, just find something that is wide and capable of making healthy electrical contacts. In fact, if you have a few lengths of old coaxial cable lying around, you do not even need to purchase costly copper braids. The cable should not be corroded or waterlogged. Remove the outer plastic jacket off the cable as gently as you can. Then push at the ends of the braid sleeve. As they come together, they loosen up, albeit slightly. Next, remove the braid from the center insulation. Now your cable is ready for removing, flattening and using for grounding RF signals. There may be some portion of the wire remaining. Use it as a high voltage piece whenever needed. Alternatively, it should come in handy for power supply applications.

Experienced people suggest that you have your ham radio station in the basement or on the first floor of your building. This proves helpful for grounding, as the ground lead should not be overlong. A lengthy ground lead makes it difficult to prevent the RF voltage from touching keys, equipment cabinets and microphones. Similarly, if your shack is on the second or third floor, your fingers will experience tingling sensations whenever they meet the metallic parts of your equipment while they are engaged in transmission duties. This is because unwanted radio frequency has energy, which flows on the leads and cabinets, instead of moving towards the ground via its intended path. Even when you touch a seemingly harmless lamp, your fingers are bound to tingle. No one likes to experience sudden shocks, whether it is a member of your family or self!

Something else that can happen is that the keyer may misfire due to the presence of RF in the vicinity. As a result, the receiver at the other end of the radio will hear only gibberish! The transmitter and frequency may become unstable, especially if the RF is very high. Thus, you should ensure a short ground system with large ground conductors.

Note that a ground that is perfectly capable of confronting lightning strikes may not be equally good for radio frequencies. It is good for providing safety to equipment, though. Sometimes you need to shield the earth from radio frequency fields while going in for RF grounding. An alternative is to move the energy generated by radio frequencies out of the lossy soil. Lightning seeks a connection to the earth and the movement of energy in/out of lossy soil. In contrast, specific RF ground applications desire the minimizing of RF current that flows into lossy soil. Thus, you will have to figure out your choices and decide what will work best for your ham radio station.

Basic Ground System

Why are we harping so much on grounding?

It is imperative that you become acquainted with the basic functions that a ground system performs.

- It is responsible for diverting or dispersing the tremendous display of energy that comes into play during a lightning strike.

- The basic grounding system takes charge of providing safety whenever a fault or problem causes the energizing of equipment or cabinet chassis. This is because it causes the appearance of dangerous voltages.

- Whenever there are end-fed (single-wire feed) antennas around, the ground system helps to provide a well-controlled radio frequency return path for them. It performs the same function for transmission-line fed antennas with inappropriate designs and poorly configured antennas too.

- Sometimes there are directly-coupled radio frequency currents flowing around. Sometimes there are induced RF currents floating around. In both cases, it is important to have a highly conductive pathway in place. The basic ground system takes care of this issue. It does not allow the currents to flow into lossy soil.

Such safety precautions are important since a ground will not aid in reception in the normal way. In case the ground does help in reception activities, it is only when there is a problem with the antenna design system or installation. This problem causes the antenna system to become very sensitive to common-mode feed-line currents. Then again, in case the addition of a station ground aids in transmission or reception, you may conclude that there is definitely a flaw in the antenna system. However, this is not to say that the ground system will never come to the aid of an antenna system facing unhealthy common-mode current issues. These issues occur because of a fault in the installation or design of the antenna or feed line. In such a scenario, the ground will ensure that the common-mode noise does not reach the antenna in its full strength. After all, the flaw lies in the antenna system, and not in the reflection of signals. If you establish the system properly, nothing that you carry out on the grounds with your station equipment will affect it. The only exception to this case is if you bring a feed antenna, comprised of a single wire, into your operating room.

It is necessary to place a ground radial system, ground screen or counterpoise right below the antenna. This will help to reduce the antenna's reaction to noise in the vicinity, thereby bringing about a reduction in sensitivity to local noise too. It is important that you choose a polarized antenna, which you may place horizontally. Earth losses are responsible for the tilting of waves and initiating vertical responses. When a signal is vertically polarized, it undergoes propagation along the earth. However, the attenuation is much less than what it would be with signals that were horizontally polarized. Ground rods prove useless in this case. Instead, it is imperative that something forms a cover for the lossy earth lying under the antenna that is horizontally polarized.

Thirdly, the ground will not help in reducing the chances of a lightning strike occurring in the region during a severe thunderstorm. Similarly, it will not reduce the number of lightning strikes either. However, it is possible to eliminate or reduce the damage that comes into play during lightning strikes, via the healthy installation and bonding of the entrance ground.

Every function of the basic ground system is unique, but it is possible that diverse functions will need to come into play at different areas in the same system. Sometimes, a single basic ground system suffices to efficiently take charge of a couple or more functions.

Safety Grounds

If your radio or radios perform their operations with the aid of power mains, there is every possibility of the power line faulting, albeit accidentally, to the chassis. The chassis refers to the metal mounting that supports the circuit components of any electronic device. Even worse is the situation wherein a linear amplifier, which has a link to a power transformer with high voltage power, suffers a secondary to primary short. In this case, the short may result in the chassis rising to both peak secondary and peak primary voltages.

To illustrate, an amplifier might have a connection to an RMS transformer of 2,400 volts. It might be operating on 120 or 240 U.S. power mains. The chassis voltage might be as much as 3,600 volts, thanks to a secondary-to-primary failure occurring within the transformer. This example should suffice to let you know how foolish it would be to reach the conclusion that your equipment does not require a safety ground connection. Every kind of amateur radio gear working in connection with the power mains, specifically devices containing high voltage, needs a safety ground.

You need not worry about obtaining a ground rod for the safety ground. All that you need is an extremely heavy conductor, which will help in setting up a connection from inside to the power mains service entrance ground. If you want more safety, you may augment this connection via some extra earthing ground rods. Similar to the grounds susceptible to lightning strikes, a special safety ground must have a good bonding with the utility entrance ground. In actuality, a reliable lightning entrance ground might be useful as a safety ground too. However, the equipment must remain bonded into the ground. Additionally, there must be a link between the entrance point panel and the operating area.

You may not use the ground lead to go beyond this barrier at the entrance panel. It could be that you have taken the precaution of installing safety ground plugs appropriately for your equipment, as well as maintained the plugs and wires up to the present codes. In such a scenario, you will not have to buy a special safety ground connection. You may do the grounding with the aid of house wiring itself.

Ground for RF Return Path

If your antenna system were to favor balanced or coaxial lines, which were appropriately installed, as well as connected, you would never have to worry about the necessity of having an RF station ground. After all, whatever currents flowed out of the antenna system would find their matches in the equal-intensity currents returning on the second conductor! This second conductor could be the identical copy of a balanced line, or a shield.

Unfortunately, the reality is far from different! You will discover that the feed systems or feed points linked to several antennas exhibit poor designs. The number of poorly-designed antennas far outweighs the well-designed ones. Hence, they keep causing issues, mostly due to flawed reasoning. The idea behind creating such antennas is that RF grounds, linked to good ham radio shacks, are there in order to lessen TVi, improve the transmitting or receiving ability or prevent radio frequency from occurring within the shack. These common-mode currents, as operators call them, refuse to behave like normal push-pull transmission line currents. You may see such currents in two-conductor transmission lines. These are the ladder or coaxial lines. To illustrate, you might have seen verticals with sparse/poor grounds. These verticals may be half-wave types too, with no radial or small systems. The verticals are antennas, which are responsible for producing extremes of RF in-shack ground currents. Similarly, you may see end-fed half waves, which include all types of end-fed dipoles. Thus, if you have always used some kind of two-conductor feed line, which results in RF in-the-shack issues, you may be sure that your antenna feed system is improperly designed/constructed. This should not happen at all, unless you have brought a long wire or a single-wire feeder directly into the shack.

It is imperative that the RF return path ground or antenna system ground have low RF impedance. This means that the ground should be able to disseminate the current around quite a spacious area. If a good-quality system is in place, it will have several conductors attached to it. These conductors, with their small diameters, will radiate out around a 1/8th wave. Sometimes the wave will radiate even further from the central connection point. It helps that these conductors have no need to encounter earth; all they do is provide electrical mass. In other words, they give rise to low RF impedance, enabling the antenna system to offer resistance to it. Thus, there is no question of capacitance or surface area at all. Instead, it is a question of ensuring that the charges are distributed efficiently over a large (in alignment with operating wavelength) spatial area.

Induced Ground Currents

Whatever kind of efficient antenna comes into play for operating a ham radio station, understand that extremely strong magnetic and electric fields surround it. Radiation is an outcome of electrical power, as well as diverse forms of fabricated and natural lightning. As for the antenna, it could be a beam, loop, vertical, dipole, etc., type. It does not matter. In case its operating wavelength is closer to the earth than to the air, you will be able to witness a large quantity of current flowing into lossy soil. When this happens, there is bound to be a loss of power. It makes no difference that the antenna and/or ground system has no direct link to the earth. The problem escalates if you place a dipole antenna at small fractions of a wavelength above the earth. It also escalates if you have a vertical antenna near the surface of the earth or mounted on it.

In order to prevent this, you may use a number of conductors spread over a large region of the earth. These conductors lie closely spaced around the antenna. It does not matter what the length of each conductor is, as long as it is able to go beyond the region where the density of the field is high. In case the conductors remain less than .05 wavelengths apart from one another, you may view them as a large and single conductor. This lone conductor is in charge of the entire region.

If the spacing goes beyond the .05 wavelength, the lossy media present in between the conductors will suffer exposure to strong fields. With regard to the gauge/diameter of grounding system conductors, it is not of much importance. What is important is the spacing between these grounding conductors, as well as their overall length.

Regardless of all these precautions, there is still one best way to keep your transmitters and receivers safe from electrical disasters. Just ensure that you have a healthy ground system and a cable entrance fed through, which has a bonding with the main ground. Secondly, disconnect the feed line from the equipment as soon as disaster is about to strike, or strikes. Take care of the disconnection in an appropriate manner. It would be good to have the feed line as far away from your operating desk as possible. In fact, it should be closer to the entrance of the shack. Then again, if you are fond of disconnect switches or relays, opt for the double-make double-break type. They will not permit lightning to pass from side to side, since the shorting bar remains grounded in the absence of energizing. Whatever arcs occur across contacts, they will enter the ground.

RF Exposure

Of late, there has been a lot of discussion around whether electromagnetic radiations are harmful to health and carry safety hazards. RF (both cell phone and shortwave frequencies) and power line frequencies have been studied, but experts have not found a link between exposure to electromagnetic radiations and health/safety hazards, and this includes the frequencies and bands dedicated to amateurs. However, one shouldn't mistake RF radiation with ionizing radiation, as the energy possessed by the signals at radio frequencies is way too low for an electron to be ionized. This is the reason it cannot lead to genetic issues, and hence, at such low frequencies, RF energy is not ionizing.

Even although there is no evidence to support RF fields posing health hazards, it is good to avoid exposure to high levels of this radiation. Considering this fact, the FCC has set limits on the exposure and there is maximum permissible exposure (MPE) set for any transmission that takes place from the radio. To adhere to these rules and regulations, amateurs should always check their respective stations to determine if their operations are causing this MPE limit to change. The RF energy only causes damage to the human body if excess power is absorbed due to the combination of power and frequency. The only risk that can be seen from exposure of the energy is thermal (which causes heating). However, no other effect (athermal) has been demonstrated at power and frequency levels used by amateurs. Quantifiable heating takes place only due to fields that emerge from the areas close to the body or strong fields.

These are two safety techniques used to ensure that people are not unduly exposed to radiation:

- Ensure strong fields are not created in an area where people are present.
- Prevent access to locations where fields are strong.

The burns caused due to coming in contact with high-voltage conducting are painful, but they're not usually life-threatening. And they can be easily eliminated with the help of effective grounding techniques or by preventing the use of antenna. When the body absorbs RF energy, it leads to heating due to the exposure to RF fields. This absorption happens because the molecules vibrate at the same frequency when exposed to RF energy. And as the RF field gets stronger, more and more molecules vibrate, resulting in more heating of the tissues. At certain frequencies, the absorption is more compared to others as the body absorbs more RF energy at these specific frequencies. So it can be concluded that the total amount of heat produced varies with the frequency and intensity of the RF field. This is known as the specific absorption rate or SAR.

The intensity of the RF field is known as power density, which is defined as the amount of energy per unit of area. Generally, power density is determined in terms of milliwatts per square centimeter (mW/cm^2); however, it is known to be the highest in the vicinity of antennas and in the direction of antennas where they have the most gain. It can also be high in the areas inside transmitting equipment. As the transmitting power increases, the power density also increases in the same degree as that of power in the areas around the antenna. When the distance from the antenna increases, the power density is reduced in proportion to the square of distance from the antenna. For instance, when the increases from antenna increase two times, the power density reduces four times. Therefore, by varying these two factors—distance and power, the RF safety of the amateur can be modulated.

The FCC has established the safety limits for RF exposure based on the known health and safety hazards, known as maximum permissible exposure (MPE) levels. As the absorption rate varies with frequency, even these maximum permissible levels change, however, since the levels are inversely proportional. So, when the absorption rate (SAR) increases, the MPE reduces. SAR, in turn, depends on the size of the body or the part that is affected. It is highest around the areas that are in resonance. For instance, if the person is grounded, his body is resonant at around 35 MHz and if he is not, it is resonant at 70 MHz. The body responds less and less to RF energy above and below the range where highest absorption happens. And the highest SAR happens in the frequency range of 30 to 1500 MHz. This is exactly how an antenna works—it doesn't respond well to signals that are coming from a source that is far from its natural resonance frequency.

The effect of RF exposure depends on the amount of heating, which happens over many seconds, and the maximum permissible limits depend on the averages, and not peak exposure. This causes the exposure to be averaged over time intervals.

- For a controlled environment, where people are aware of their exposure, the averaging period is six minutes.
- For an uncontrolled environment, where people are not aware of their exposure, the averaging period is 30 minutes

This variance in the averaging period shows the difference between the time when people are present and exposed. People tend to stay less in uncontrolled environments and therefore the averaging period is longer. During this averaging period, a transmitter might only generate RF for a certain amount of time. The transmitter output for most contacts made is about 50 percent of the time. Due to this pattern, the emission duty cycle lowers. A duty cycle is the ratio of on-the-air time of transmitted signal to the total operating time at the time of measurement. It can also be defined as the percentage of time when a transmitter transmits.

The duty cycle impacts the average power of transmissions and should be taken into account at the time of exposure. It is a simple calculation—the lower the duty cycle, the higher is the output from the transmitter, while still maintaining the average value within the limits. As most amateur operations are irregular, the total time spend on transmitting on the air is quite low. For instance—when a roundtable is happening among three hams, each of them is likely to transmit just 33 percent of the total time. This further reduces the exposure, irrespective of the mode being used.

Emissions with lower operating duty cycles, for a given peak envelope power (PEP), create less exposure. When this PEP is multiplied by the operating duty cycle and time spent in transmitting, what we get as the output is the overall average power for that period.

Overall Average Power = PEP x Operating Duty Cycle x (time taken in transmission/averaging period)

Here, there is one more effect that should be taken into consideration—the effect of antenna gain. Beam antennas create gain by focusing the radiated power in one direction. This gain helps in increasing the average power in the given direction, thereby decreasing the power in other directions. So there are four factors that impact the exposure, namely frequency, transmitter power, radiation pattern of the antenna and the distance to the antenna. If the antenna that is being used has an associated gain, the effect of this gain has to be taken into consideration for determining the exposure. For example, if the antenna that is being used has 3 dBi of gain, which corresponds to a four-fold increase in radiated power in the preferred direction, to determine the RF exposure in the forward direction of antenna, the average power needs to be multiplied by four.

As per the rules set by the FCC, all stations that are located at a fixed location should calculate the RF exposure. This is because mobile and handheld transceivers are exceptions. This evaluation can be done in three different ways; however, the most commonly used method is the one that uses simple formulas and tables to determine whether a given station can cause an exposure hazard. The power density of the transmissions can also be measured using an RF power density instrument. If not, one can create a computer model of the station and use the results to determine the value. However, neither of these methods are commonly used due to the effort involved.

Before the evaluation is performed, it is good to check whether your station is on the list of exemptions. A station is exempted from evaluation if the power it transmits to the antenna is less than the required levels. However, once the evaluation is done, there is no need to recalculate the value unless the equipment is being changed in the station. This is because each piece of equipment contributes towards the average output power, for instance—it impacts the transmitter power or the gain. Even when a new frequency is added to the band, there is a need to reevaluate.

To do the evaluation, we need information around the signal's power level, its frequency, distance from antennas and its radiation pattern. The following steps will help you with such an evaluation:

- Take the average power of each transmitter on the bands. Starting with the PEP, apply the corrections for mode and patterns.
- Subtract the feed line losses if you have been given a long coaxial feed line.
- Include the impact of antenna height and gain with the help of given ARRL tables.
- Determine the distance from antennas to comply with the given limits of MPE.

The evaluation should be done at each frequency band and the antenna used on that band. The task can be made easier by doing the evaluation at the highest average power, usage and mode of the antenna as this will give the minimum requirements for all conditions for that specific frequency and antenna. Once the evaluation is done, compare the results of minimum separation with the actual value of installation and you will see that there are mostly no health hazards. This means that most of the stations do not cause any health hazards.

But in case you do find a potential health hazard, or if you are just beginning to build the station and wish to prevent all health hazards, here are some tips you can follow:

- If you are using a beam antenna, do not point it in the direction where you are likely to find people.
- Antennas should be located in regions where there are few people as this reduces the chances of them touching an energized antenna. Even if it has low power, an antenna can cause RF burns.
- The average power of the transmission should be limited by transmitting for short periods of time or by using lowering the duty cycle mode.
- It is good to raise the antenna as it also improves the signal strength in distant locations.

- Use an antenna with lower gain as it reduces the radiated power density or the transmitted power. You can also make contacts with lower gain or power.

All these techniques will reduce the exposure to you and others around you. This way you will find a combination that offers minimum effect, while still ensuring you still are within the permission MPE limits. Even when the emissions are made using handheld or mobile transmitters that are exempted from evaluation, there are certain ways that can be utilized to minimize RF exposure:

- The mobile antenna should be placed on a raised platform. For instance, the roof or trunk of a car, to minimize the shielding effect.
- A remote microphone should be used to hold a handheld transceiver away from the hand while it transmits.

Décor

True, you are into the serious business of operating a ham radio! True, your shack is just a house for all your equipment and your operating desk. However, there is no harm in helping it acquire a professional appearance, is there? This is something that many ham radio operators do not place any importance on, but you may opt to be different! You are the ambassador of your hobby. Therefore, strive to be an inspiring ambassador! Do not allow your shack to resemble a pigsty, be disorderly, etc. Whoever walks in should just go "wow!"

To begin with, let any visitor receive an eyeful of your awards and certificates, immediately upon entry! Are you wondering how to obtain them? Well, that is easy! Several organizations and clubs love to distribute certificates to members. Therefore, you can have several membership certificates for display. Then again, you may strive to win all manner of awards by participating in competitions. Whenever you win awards or award certificates, let people see them so they can comprehend the seriousness with which you tackle your hobby. At the same time, do not just tape them to the walls of your shack or place them on shelves. Have a good-looking but simple frame for every certificate. Peruse local shops which specialize in making frames for certificates. They should be able to offer you moderately-sized frames at reasonable prices. Sometimes the price comes down when you buy in bulk. At times, there are clearance sales too, with products going on sale at half their original prices.

Just in case you wish to be innovative, you may choose to purchase a number of frames, all in odd sizes, but displaying the same style. They will give your shack décor a unique look! It could be that you like certain frames, but they seem rather huge in size. Well, have them customized for your certificates. If stores have the capabilities to fulfill your request, they will do so. Note that the majority of frames have no mounting kits nailed to their backs. Therefore, you will have to purchase several mounting kits too, from the same shops that sold you the frames. While attaching a mounting kit to a frame, remove the frame's glass first. Otherwise, your vigorous hammering may break the glass. If you do damage a few glass sheets, opt for pre-cut plexiglass sheets as replacements.

All right, what else can you display in your shack, along with your membership certificates and award certificates? You can show off your current license, indicating to everyone that you are an official ham radio operator. It does not matter that you are an amateur and not an expert. After all, the license comes into play every time you are on the air. In case you have several licenses, old and new, just paste them in attractive formats on black construction paper prior to mounting them. This way, your visitors will even get an idea of the history behind your ham radio hobby! It could be that space eventually becomes a concern with so many certificates, licenses and awards. You can then place older certificates behind the newer ones. It will be akin to a 'cycling' display! Even if you want to give your shack a complete makeover in the future, all you have to do is to shuffle around the certificates, using the same frames and mounting kits.

Should the walls of your shack display only certificates and licenses, or should you use them for highlighting something else too? Well, if you find that you have very few certificates and licenses, since you are just a beginner in this arena, your walls are bound to look rather bare. See if you can make photocopies of schematics to radios or old radio advertisements. After all, there must be things that you wish you owned! Use dark backgrounds. You may use a tan background too. After copying the requisite images, trim the paper to fit various frames. See if you have small picture frames lying around your house, which are of no use for anything else. When they are ready, arrange them on your walls.

Another idea is to display pins (pieces of jewelry that go on garments) that are just lying idle in your dresser. Towards this end, obtain a healthy-sized frame. Remove the glass and view its shape. It should help you to cut out a piece of corkboard in a similar shape. Use three to four coats of spray paint on this corkboard. This is the background for your pins. Keep adding more pins as you go along. If you should need any of them for wearing to an event or convention, remove the requisite pins. Replace them on the board soon after you return home. However, keep every fastener safe. You may not realize it, but the pins make for an impressive wall hanging!

Apart from this, you may deck the walls of your shack in other creative ways. For instance, have pictures of your family, close friends, activities that you have participated in before, etc. World maps are also welcome, for you are dealing with the entire world via your ham radio. Have framed postcards of all the places you have been around the globe. In case you can obtain a dry eraser board, use it for tracking information regarding reports, your points gained via participation in contests, etc. A shadow box filled with important items from your ham radio past is also a great idea. These are but a few examples of how you can improve the look of your shack. You are welcome to come up with your own ideas.

Upgrade Your Shack

Although you may be living in the jet age, you may not try to maneuver your ham radio station in the same way! For instance, soon after you go through your first QSO, you may begin to wonder if you should not be concentrating on upgrading your ham radio shack. After all, you wish to be stronger or to be able to hear better. However, you need to know what to upgrade, and how to upgrade it. Therefore, do not just give in to impulses, but read the tips outlined below!

In case you are truly facing issues with transmission and reception, you should be working towards obtaining better-functioning antennas. However, do not just squander money on the most expensive ones that you see, believing that they will give you better value for your money. Instead, opt for raising your antennas, and then, make them larger.

Similarly, if you feel that your station should be having more power, wait until you have upgraded your antennas first. You must ensure that you are able to hear well, prior to extending the range at which you feel that others will be able to hear you better, too.

Do not give in to the urge to obtain an amplifier. It will not help you hear better in the least!

Do not give in to the urge to purchase a brand-new radio either. Instead, attach additional receiving filters to your existing one. This will prove cheaper in the end. It could even be that you have not been operating your old radio properly. Maybe you lack sufficient knowledge. Therefore, upgrade your ham radio education. This will be the best decision you have ever made!

Since you are not able to hear properly, fix the multimode processor between your ears. Obtain a better one.

To conclude, you must upgrade with an open mind and the patience to take things one step at a time. At the same time, keep yourself updated about all the latest developments on the ham radio scene. Tell yourself that you will read, listen and discuss things throughout your ham radio journey. When you improve your comprehension and capabilities, you will be able to fulfill your operating goals all that much faster. Towards this end, request advice and guidance from experienced operators, members at your club and so on.

Chapter 8—Let Your Radio Announce Your Airy Presence

It is time to get on the air now! Towards this end, you will need to consult reliable experts on how to set up your workstation, select the best possible radio equipment and antenna, pick the right computers for your shack and so on. With regard to equipment, you may buy used materials and appliances, provided they serve you well. In short, you will have to gather valuable information before engaging in purchases to launch a functioning radio station.

A. Goals for Setting up the Radio Station and Usage of Personal Resources

Working from Home

Would you like to build your business at home? Well, then choose a semi-permanent installation, which should serve beautifully as your home station. You will not need much space and furniture for setting up the radio equipment, thereby ensuring that the rest of the family remains undisturbed by your activities. In fact, you may even opt for a shack in the basement, provided it is not located directly under a bedroom. This is because you will be speaking loudly and listening to voices from the other end on a speaker. Attics and garages do not serve the purpose either, since they are inclined towards wide fluctuations in temperature. Therefore, select a basement that is dry, or a spare bedroom. Wherever you choose to be, your headphones must work splendidly. Purchase high-quality ones.

Another important aspect of your amateur radio services are the external antennas. They must have great feed lines attached to them, in order to transmit clear signals. You will need to ensure that the equipment for holding them up is strong and safe. Should you decide to use rotatable beams, mounted on towers/masts, you may require official approval. Do not hesitate in this regard, for the main purpose of setting up a ham radio station (despite its amateurism), is to provide assistance during emergencies. It has to work, even if the AC power goes off. In case your radio is compatible with 12 volts of electricity, you will be able to put your car battery to good use, at least for some time. Even computing gear and other accessories require power. A generator should suffice, if linked appropriately to the radio station.

Going Mobile

Nowadays, it is easy to obtain radios that are small, yet multi-modal, and all-band in nature. With their help, you should be able to have easy access to HF, UHF and VHF bands. You need not spend too much money on this equipment. In contrast, you may have to spend more on putting an HF mobile antenna in place. You may not crab about the budget, because you are aiming for a customized installation! This will require a unique set of considerations. Then again, although it should not be too difficult to figure out the automotive wiring and fixtures all by yourself, it would still be advisable to consult a professional. After all, safety is also of prime importance. You will just have to shell out a few more dollars. Alternatively, you may peruse genuine websites and articles that focus on mobile operations. One such website is K2BJ.

Going Portable

It is possible to opt for a portable operation too, considering that you have the good fortune of selecting any level of portability, from minimalist backpacking to actually positioning the RV somewhere. You should have the capability of packing or carrying your complete radio station, along with its power source, wherever you go. Therefore, allot a total weight budget to your business venture too, wherein you may use your imagination to obtain the best antennas and accessories. They should suffice to grant you better options for your radio and power sources.

It may surprise you to view the awesomely small radios that you find in the marketplace. However, you may not find every single radio so easy to operate, especially if you are a novice in this field. In such a scenario, you may want to give up on small radios, and instead opt for a rig with more features, which is easier to handle. Additionally, you will gain good operational practice. With experience, you will also be able to figure out what features you may discard. To illustrate, if the operating/station time works on a limit, it might best to focus your attention on a lone band, rather than on multiple bands. Other portable operators and low-power operators generally favor 14, 17 and 21 MHz on HF. It is possible to keep these frequencies open for the major part of any day, since the antennas are small and portable. In case you prefer operating at might, go for seven and 10 MHz. With regard to VHF, you may opt for 50 and 144 MHz. There are several operators on these bands, since they offer splendid propagation.

Tackling Handheld Radios

It is handy to have a handheld VHF/UHF radio with you, since it helps you to keep in easy touch with your family members and friends, albeit in the local vicinities. At the same time, such radios prove highly useful during personal or club outings. This is because these handheld radios often cover a large receipt area. As a result, you gain access to bands from the fire and police departments, as well as commercial broadcasts.

Now, are you buying your very first handheld radio? If so, choose a simple model, featuring a single band. After gaining some experience, you will be able to opt for a model with multiple bands, possessing all the bells and whistles. Then again, as a novice to the arena, you will find simple radios even easier to operate, since they display all the features that should prove useful for your business. In case you decide to go for an older and used UHF/VHF model, which may be mobile or handheld in nature, ensure that there are built-in sub-audible tone features too. Your radio may require some modifications or accessory boards in order to include the sub-audible tones. This is an important prerequisite. Therefore, think twice before buying an older rig. You may find it difficult to find sufficient parts to initiate any required modifications.

Moving on to the accessories—are they important? Yes, they are, for they extend the usefulness, as well as the life, of a portable radio. Here are some useful accessories.

- You should be able to extend the range of your handheld radio, even at home, via an external, full-sized, metal antenna. This is because you may not find the rubber duck flexible antenna very useful. This is only good for the portability of your station.

- It would be good to opt for a low-loss feed line, which is still of high quality, for cables that go beyond much more than a dozen feet.

- You may not feel comfortable holding the radio to your face all the time. Therefore, purchase a speaker-mic combination.

- Have a jacket or case in place to take care of your radio while you put it through the rough and tumble of carrying it around.

- Do not ever go anywhere without a few spare batteries on hand. In case you have a rechargeable battery, ensure that there is a spare charged one in place too. While the wall transformer model is great, you would do better to opt for a drop-in charger. The latter charges batteries much faster than the former does. If you are lucky, the manufacturer may even offer a battery pack which accepts ordinary AA cells. Whenever you are not able to use the charger during an emergency, use this battery pack.

- Always keep track of the model and serial numbers displayed on your radio. Ensure that the case displays your name and driver's license number too. This will prove useful in out-of-the-way places, where portable and mobile radios tend to become easily lost. Sometimes people steal them, too! Then again, you never know, but you may have to utilize a larger radio even on portable expeditions. Therefore, insure your equipment against loss and theft.

- It is important to know about your auto insurance and homeowner's insurance too, since they have links to your radio equipment.

As discussed earlier, you need to gather all kinds of items, apart from the radio itself, when launching a radio station. They include cables, antennas, power sources and accessories. You should be able to figure out how much you will spend on each item by allocating a specific budget to various resources.

B. Tips for Selecting a Radio

It makes no difference what kind of radio station you are planning to launch. The new radio will make a big dent in your budget! Nonetheless, you should not mind too much, for this radio is the foundation of your ham radio venture. After all, you will be interacting with it more than with anything else. Therefore, you need a good performer.

Shortwave/HF Radios

There are three categories of HF radios: basic, Journeyman and high-performance radios. Although modern radios are perfectly capable of receiving and transmitting at wonderful levels, they still exhibit differences in certain areas. Some of these areas include:

- the capacity to receive even when strong signals are around
- the ability to receive signal filtering
- the capability of controlling filters
- possessing built-in antenna tuners
- covering one or more UHF/VHF bands
- possessing operating facilities (number of memories, sub-receivers)

Basic HF Radio

Such a radio possesses a simple set of controls. It includes signal adjustments and a basic receiver filter. You may retain a fixed value for the controls, along with on-off settings. The metering and displays of this radio are rather limited in nature. Then again, you may connect it to a single antenna, expecting minimal support for your external accessories. Nonetheless, if you are a beginner in this area, you will do well to obtain a basic HF radio. Later on, you may utilize it as a portable or a second radio. Sometimes you may have access to a computer interface too.

Journeyman

Since it is a step higher than the basic radio, it offers you all the vital transmit and receive adjustments. The majority of radios belonging to this category have front-panel controls. You are bound to be happy at the sight of existing functions linked to metering, and an expanded set of memory and display. Sometimes you may come across Journeyman models possessing extra bands and offering support for digital data operations. Things like internal antenna tuners and connections for external equipment are common. The connections for external equipment include band-switching gear and transverters. A computer should be able to take charge of the available computer interface.

High Performance

The front panel displays an extensive set of controls which are capable of receiving and transmitting signals. It is possible to configure these controls under a menu system. Apart from this, you should be able to find a state-of-the-art receiver, as well as a sub-receiver. Both of them have complete interfaces for computer control and digital data. The manufacturers also provide standard internal antenna tuners, as well as some kind of antenna switching. Do not be surprised to find computer-style displays, as well as complete metering and displays, for they are popular in these kinds of radios.

Select a Filter

A receiver needs to use filters in order to prevent signals from nearby vicinities from interfering with a particular and desired signal. It is the responsibility of these filters to reduce the strength of unwanted signals or attenuate them as soon as they reach an area just a couple of hundred hertz away from the station. Earlier, only quartz crystals could do the job of behaving like filter components. Therefore, they became famous as crystal filters. However, other filters have entered the marketplace since then. They are mechanical filters, which use vibrating discs, despite being similar in appearance to quartz crystals. These receiving filters take charge of the radio signals before they're even converted to audio. Towards this end, they function on a fixed bandwidth and refuse to undergo adjustments. Bandwidth refers to the range of frequencies that the filters can pass. You may discover the frequency of each signal in hertz.

Since the bandwidth is fixed, you cannot use the same filter for all radio stations. Therefore, you may peruse various filters with diverse bandwidths. Regardless, whenever you receive your radio, it should have at least a single sideband or SSB filter in place. This is a commonplace filter which is a feature of high-frequency radios. Alternatively, you may find an FM filter, which is a common feature in mobile or handheld instruments with very high frequency. During the initial stages, you may choose a standard filter bandwidth, which is suitable for HF SSB operations. This bandwidth is 2.4 kHz. In case you are operating your radio station in a crowded environment, wherein you incur some loss of fidelity too, you may opt for filters possessing bandwidths of 1.5 to 2.0 kHz. In case you need filters for digital data and Morse code, use standard filters, such as 500 Hz wide. In fact, this is a good option. Narrower filters are content with 250 Hz. In fact, the most popular filter is the 500 Hz CW filter.

The next option is the narrow SSB filter. Sometimes, certain high performance and Journeyman signal receivers have filters that love to cascade. This means that one filter follows the other in function, akin to the water in a waterfall. This cascading happens at more than a single intermediate stage. Therefore, if you do not mind spending a little extra, you may purchase extra filters. After all, they can make a significant difference to your operations, specifically in the ability to receive signals on a crowded band.

What is digital signal processing (DSP)?

Here, the P may also refer to processor. This is a microprocessor within the radio which functions on specially installed software. The idea is to process or operate on incoming signals, such as audio signals, prior to their becoming amplified for output over the headphones or speakers. If your DSP is highly advanced, it can take charge of various signals at radio frequencies. In fact, it is capable of undertaking filtering functions that can remove signals that are off frequency. It can also eliminate or reduce several diverse types of noise. It can automatically detect and destroy interfering tones. As a result, you should find DSP filtering a more flexible option in comparison to that performed by a crystal filter. At the same time, you may view the performance by DSP as akin to a healthy fixed-width crystal filter. In other words, if your DSP performs at higher frequencies and takes charge of higher numbers of bits specified for it, it indicates a healthier processing performance. You should find all information about it in the operating manual for your radio. Even the product specification should supply sufficient information about your DSP.

HF Radios for Mobile and Portable Operators

It does not matter whether you want to be a portable or mobile operator. A radio that is eminently suitable for your particular style of operations is available, thanks to annual improvements. Even the smallest of radios exhibits better features and bands, thereby permitting ham operators to function in small or cramped spaces. They are perfect for limited spaces at home or conditions where dual portable/home stations are in place, capable of covering both types of bands, UHF and VHF. At the same time, bear in mind that if your rig is small, it will make compromises in certain ways, in comparison to rigs that display designs for high performances. As it is, your operator interface is compulsorily menu-driven in nature. In turn, this makes certain adjustments less convenient in nature. Of course, you must understand that the commonly used controls need to remain on the front panel all the time. What you will have to compromise on are internal antenna tuners, each possessing an output of 100 watts. Only high-class rigs can have them.

Now, if you would like to fix a radio in your boat or vehicle, where do you place it? If you own a yacht or RV, you should find this to be no great problem. However, if you have an 18-foot runabout or a compact car, you will have to deal with the space issue as best as you possibly can. Fortunately, radios that fit into mobiles possess detachable front panels. These panels are famous as control heads. You may use the same for your limited spaces. You should find it easy to place the body of your radio on a bulkhead or a seat, or even under the dash. After all, the body and panel can separate easily. In case you share your boat or car with someone else, come to a mutual agreement about the exact locations for drilling holes, if necessary. Similar to what you would consider for a larger base station, you will have to take note of several aspects and accessories before you pick out a radio suitable for your purpose. You may even request a trial performance before you actually purchase one.

Digital Data for High Frequencies

HF radios are going in for the provision of a connector or a couple of connectors in a big way! These connectors are comprised of digital data interfaces which help in bringing about easier bonding between a personal computer and conducting operations on digital modes. These digital modes could be PSK31 or RTTY. Some of these radios are comprised of inbuilt data modem or TNCs (terminal node controllers). The TNC is a kind of data modem which you find in packet radios.

Keep a lookout for accessory sockets present on radios. Radios with such sockets favor certain signals. One of them is frequency shift keying (FSK), wherein the digital signal originating at this connector pin ensures that the transmitter brings about the output of two tones suitable for frequency-shift keying. Using this method, you may transmit dual frequencies that are generally used for radio teletype or RTTY.

Then again, if your radio possesses an internal data modem, you may bring about connections between the inputs and outputs of digital data and a computer. However, you may require the services of an RS-237 converter in order to achieve the data in/out signals. RS-237 is a kind of serial communication.

Line in/out refers to the inputs and outputs of audio signals. These signals have a compatibility with the signal levels emitted by the sound card within your computer. You may use this kind of input for generating digital data whenever the sound card of your computer behaves like a data modem.

With regard to PTT, it is akin to the push-to-talk button present on a microphone. With this, you will be able to key your transmitter via external equipment or your computer.

DISC, or discriminator, refers to a specific input. This input is actually the unfiltered output that the FM demodulator generates. If you have external equipment for this purpose, you will find it useful to utilize this signal for both, tuning the indicator and receiving data.

You may not be able to configure your radio in such a way that it supports digital data. However, the manufacturer's website may be able to help you with relevant instructions. Alternatively, you may request that a dealer supply you with a detailed radio manual. Ensure that you find all the information related to setting up appropriate connections for packet radio, radio teletype operation, garnering digital data, etc., in the manual. You should be able to figure out if your favorite mode receives the appropriate support or not. In case you find that the manual is not of much use to you, approach the concerned manufacturer directly. There should be someone who will provide guidance on hooking up your radio. The last resort is to peruse digital data websites. They suggest how you may interface various radios to your personal computer through a variety of files.

Deciding on Amplification

Until you gain some experience of being on the air, you would do well to avoid purchasing an amplifier for your high-frequency operations. You need some kind of expertise to handle an amplifier and the issues associated with it. These include safety of radio frequencies, power, interference and feed lines. Additionally, the amplifier magnifies the radio signals, making them stronger. This additional strength has an effect on more hams, specifically, if you have not used the amplifier properly or not adjusted it properly. It is akin to learning to drive, slowly and surely!

Note that the majority of high-frequency radios generate an output of 100 watts or beyond, compelling you to conduct heavy operations at any time. For instance, you will have to figure out when your output should be 500-800 watts, or the complete 1,500 watts. Since the extra punch suffices to get the job done, DXers prefer to opt for them while making contacts on difficult bands, possessing long pathways. For instance, sometimes a traffic handler finds that the band is too noisy or crowded. In such a scenario, the amplifier helps him/her to read the messages coming through quite clearly by switching itself on. Digital operators find the process of amplification useful for minimizing errors found in data that they receive. Then again, whenever there are emergencies, the amplifier helps the signals to reach a ham radio station possessing damaged or poor antennas.

If you are serious about obtaining high-frequency amplifiers, you will find them in two varieties. One is in the solid state, while the other is a vacuum tube. The solid-state variety is rather complex in nature, but user-friendly. You need not opt for any warming up or tuning. Instead, just turn it on and relax! The vacuum-tube variety is marvelous for places using high levels of power. Additionally, this kind of an amplifier is more affordable per watt of output. While solid-state amplifiers are larger, vacuum-tube amplifiers are rather fragile in appearance. Never buy CB footlocker-type amplifiers, for they have deficiencies in design that you may not be aware of, which will greatly reduce the quality of your signals.

All about UHF and VHF Radios

Several radios offer high frequencies of 50, 144 and 440 MHz. In fact, some specimens offer an output of 1,200 MHz! As a result, you need not buy a second radio because you want to conduct VHF or UHF operations. You could imitate other ham radio operators, who combine a HF/UFH/VHF radio, with all the bands on it, with a UHF/VHF FM rig. This suffices to make use of the packet radio and local repeaters.

It is easy to differentiate between radios only for FM modes and those which are multi-modal or function on all modes. These UHF and VHF specimens can perform operations on FM modes, SSB or single-side bands and CW. If you want to get ahead of your competitors by favoring these modes and bands, just purchase a dedicated radio. However, the multi-mode variety of rigs is not so easy to find nowadays, since several models of all-band radios suffice for UHF or VHF coverage. Additionally, these models help you to handle amateur satellites easily. This is possibly due to the presence of automatic compensation and completes duplex operations for tackling transponder offsets. You would find it very difficult otherwise, for amateur satellites are on the move always and require cross-band operating. Their movements result in Doppler shifts on all the signals that your radio receives.

Apart from this, you may choose an all-mode radio, which is capable of functioning on amateur microwave bands. However, you will not be able to obtain commercial radios for bands such as 900 MHz, as well as GHz of 2-24, and beyond. An alternative is to buy a transverter, which will help in converting all the signals that microwave bands receive into 28, 144 and 440 MHz bands. Your radio will award these signals the same importance as it does other signals. The transverter is also capable of converting an output of 100 milliwatts and below, emanating from the back-up radio, to a band that is higher in functioning. However, you will also need an external amplifier, if you plan to drag your output to 10 watts and higher.

Radios with FM Only

Almost every ham radio operator falls in love with the FM on his/her UHF and VHF bands! It does not matter what individual mode or operating style is a favorite! In actuality, even someone with a new technician's license may prefer a handheld radio or FM mobile at the initial stages. It helps that the all-mode rigs have FM. Yet the popularity of the FM mode prompts people to set up rigs with FM only. While the handheld specimens may be single band, dual band or multi-band in functioning, the mobile rigs may behave as base stations at home too.

Surprisingly, even with the multi-band model, covering a range of 50 to 1,296 MHz, many people prefer the lower models! The reason is affordability. To illustrate, if you were to buy the single-band radio, you would have to pay less than half the price that you would pay for a multi-band radio. This is specifically true for two meters (VHF 144—148 MHz) and 70 cm bands (UHF 430—440 MHz), wherein you perform the majority of your operations. Hence, extra bands do not serve much purpose. Such a radio is bound to have certain standard features. One of them is the encoding and decoding of tones, which grant access to repeaters on a limited basis, such as CTCSS sub-audible tones. Another is a DTMF keypad, which is great for entering control tones. The control tones are the same as those that you view on a touchtone telephone. Other inclusions are a dozen or so memory channels and variable repeat offsets. You will also receive a simple charger along with a rechargeable battery.

Would you like to receive extended coverage too? It proves useful for listening to broadcast FM. Similarly, if you have low-VHF land-mobile bands associated with the two meters, you should find it extremely useful. The VHF land mobile is great for paging, business and public safety. A marvelously entertaining, as well as emergency feature, is having TV channels from two to 13. They should be able to receive anywhere between 54 and 216 MHz.

Now consider power output. If you prefer small/credit-card sized radios, they are not very powerful despite being conveniently portable. Note that if you do not have access to excellent repeater coverage in your area, you will remain out of touch even at the times when you need this feature the most. Therefore, choose a radio that offers at least a watt of output.

Mobile Radios

Some of the features displayed by mobile radios are similar to those highlighted by handheld radios. They include controls, memories and scanning. If you opt for extremely powerful transmitters, along with an external antenna, you will be able to increase your range drastically. In fact, the receivers that exist in mobile radios often exhibit improved performance in comparison to those which handheld radios display. They are capable of rejecting the strong signals which are the outcome of land-mobile dispatches. They are also capable of paging transmitters which are on adjacent frequency bands. You can learn more by perusing product reviews printed in magazines related to ham radio operations. Such reviews exist on official websites too. You may even consult members of your club about their experiences. After all, they operate in the same locations as you do.

It is possible to utilize mobile radios for conducting digital data operations on UHF or VHF bands, particularly on packet. While they had to face limits of 1,200-baud data (data transmission rate for modems in bits/seconds) earlier, ham radio operators now have the freedom to choose 9,600-baud data in modern times. Therefore, if you are planning to utilize your mobile rig for digital data operations, you must ensure that it is capable of handling such data. Furthermore, your rig must be rated for 9,600 baud without any modification.

Sometimes you may see a ham radio rig claiming that it is compatible with APRS or GPS. APRS refers to automatic position reporting system, while GPS refers to global positioning system. Whenever information about a particular location reaches a GPS receiver, it moves, via a transmission pathway through packet, to a network comprised of APRS computers. The process takes place over a VHF radio. Any radio which is compatible with APRS possesses a serial port for a GPS receiver. There is also an internal packet TNC which is capable of both sending and receiving APRS packets, even in the absence of external devices.

UHF/VHF Amplifiers

It is common practice to increase the transmission of power with the aid of a handheld or small mobile radio which is all-mode in nature. The addition of amplifiers makes it possible to transform even a minimal number of watts of input into something beyond 100 watts of output. In their solid states, we call these commercial units bricks. This is because the amplifiers are akin to large bricks displaying heat-sinking fins on the surface. If you wish to improve the performance of a handheld radio, you may buy a small amp, as well as an external antenna. That way you will not even need a separate rig for your mobile.

Amplifiers are compatible with SSB/FM models or FM-only models. When you opt to use an amplifier with an FM-only radio, you may expect serious distortions to occur even when a single-sideband signal is coming through. In contrast, the amplifier that comes into play for SSB/FM models works better, thanks to the switch that enables changing between the modes. As a result, it becomes possible to amplify signals coming through in Morse code, regardless of which mode is in play at the time. However, you will profit more if you choose the FM mode.

The linear amplifier is perfect for use by SSB models. Note that you will have to contend with more pronounced RF safety issues, especially if the MHz is beyond

30. This is because the human body has the capacity to absorb energy much more easily at such frequencies. If you use a beam antenna, you may be sure that the outputs coming from the amplifier are powerful enough to be hazardous in nature. Therefore, if you find that the antenna is too close to people, don't buy an amplifier at 50 MHz.

In conclusion, whenever you have to force yourself to choose between mobile radios or handheld radios, go through the list of features associated with each model. Check the capabilities of each selected radio. Search for features that you may want to use in future. Then again, you may like something just because it falls into the nice-to-have list. Whatever you do, choose wisely and well!

C. Tips for Selecting an Antenna

It is hard to differentiate between the importance awarded to the antenna and that awarded to the radio. After all, it is not easy to make up for the deficiencies exhibited in one by making improvements in the other! However, there is no denying that if the antenna is of good quality, it can make even the weakest of radios sound much better. The reverse is not applicable here. Therefore, it is imperative that you grant equal importance to choosing both your radio and the antenna. The focus here is on the antenna, since we finished dealing with radios in the earlier section. There are diverse types of antennas for your ham radio station. See which ones suit you best.

Antennas for High Frequency

They tend to be quite large, for the very purpose of efficacious functioning. To illustrate, a vertical antenna for 1/4-wavelength takes the appearance of a 33-foot high metal wire or tube. Its dimension is 40 meters. Should the HF frequencies go still higher, the antenna's size may come down to eight or 16 feet. Nonetheless, the antenna for a ham radio station is definitely bigger than the one utilized by a big television set. Note that your selection of the antenna is dependent upon your physical circumstances. There are all manner of designs in the marketplace in order to ensure that your on-air performance remains top notch!

Generally, ham radio operators across the globe opt for beams, wires or verticals, as far as antennas are concerned. It is your choice to purchase them from vendors selling equipment for ham radio stations, or build them yourself.

Beam Antenna

Beam is short for beam forming. It refers to an antenna that has more than an element to help it concentrate or steer its capacities for transmission and listening in a particular direction. A beam antenna may be comprised of just two wires, or could reveal a complex collection of over a dozen pieces of tubing.

To date, nothing has taken the place of Yagi's popularity. Yagi is a beam antenna that came into the public eye during the 1920s. Two Japanese scientists, Dr. Uda and Dr. Yagi, invented it, hence the name. This specimen is comprised of a couple or more wires or tubing, which lie parallel to one another, with an approximate wavelength distance of 0.1. These are the conducting elements of the antenna. Then there is the driven element, which refers to the element connecting to the feed line. Reflectors and directors, which are the other elements, help in reflecting or directing the energy in a specific direction. Manufacturers cut the reflectors and directors in alignment with certain measurements and even space them for maximum functioning. Since these elements take control of directing the energy flow without having a direct link to the feed line, they are considered parasitic elements. Sometimes to Yagi antennas are also referred to as parasitic arrays.

Of course, despite its popularity, the Yagi has undergone certain modifications over the years. Today it is comprised of a design with three elements in it: a director, a reflector and a driven element. They function on three popular ham radio bands, which are 10, 15 and 20 meters. Therefore, it is popular as a tri-bander. If created from wires, the Yagi will have a lower frequency. Such a Yagi antenna is also available in the marketplace. To illustrate, you may have a Yagi beam antenna mounted on a mast measuring 55 feet, possessing an extremely low operating frequency of 14 MHz. With regard to other beam antennas, they are comprised of triangular or square loops. Although they function on the same principle as the Yagi beam antenna, they display loops of wire, and not the straight elements created from tubing or rods.

Square-loop beam antennas are known as quads. The triangle beam antennas are delta loops. If you come across a certain type of beam antenna, which resembles an oversized television antenna with several angled elements, you will know that you are seeing a log periodic or log. Thanks to the large number of elements in the antenna, it manages to come into play on all bands, albeit within a certain range. This is similar to how television antennas work, wherein a television set is able to receive so many channels. Ham radio operators opt for logs in order to cover an array of bands, such as 10, 12, 15, 17 and 20 meters, via a lone antenna.

Unlike vertical antennas and wire antennas, beam antennas have the capacity to rotate. The former move or radiate in all horizontal directions, albeit equally. The latter manage with a fixed orientation. Designed from aluminum tubing, rotating beam antennas permit the ham radio operator to focus on a specific signal. At the same time, the operator may also reject a signal that is arriving from a certain direction and causing interference. It is possible to position small high-frequency beam antennas on rooftop tripods or masts. These supports need not be expensive at all. Most of the structures designed to support television antennas tend to undergo overloading, unlike the beam antenna.

Note that you will need a rotator, which you will have to mount on a positioned support, in order to turn the beam antenna. You may control this rotator while sitting inside your shack, for there is a meter that displays the direction. Generally, ham radio operators show an inclination towards vertical or wire antennas. They do not choose beam antennas immediately. Nonetheless, you would do well to test your hearing skills and equipment for a while. Listen to the signals coming over the air. This will help you decide which antenna will work best for your operations.

Wire Antenna

The simplest model of a wire antenna is the dipole. There are only two constituents in this model. One is a piece of wire, which displays a cut in the center. The other is the feed line, to which the cut wire bonds itself. Despite its simplicity, the dipole is a great performer. In fact, it is definitely much better than a simple antenna.

To construct this simple dipole:

First, obtain a copper wire which has a 10- to 18-gauge. You may opt for a solid or stranded piece which is either insulated or bare. When your wire is ready, it should show these measurements.

Length in feet = 468/frequency of use in MHz.

As per this formula, you should be able to witness a slight shortening effect. As a result, the 1/2-wavelength of wire appears slightly shorter in comparison to the 1/2-wavelength in air. For instance, if you obtain a dipole which is 468/21.1 = 22.2 feet long, it works on a band of 21.1 MHz. Next, ensure that there are 18 extra inches at each end. You will need these extra bits for connecting to end insulators and tuning. You will also require another foot of wire, approximately six inches by two, which should connect to the center insulator. Thus, your complete dipole will measure 22.2' + 18" + 18" + 12" = 26.2'.

Coming to the actual assembling of a dipole, follow these steps:

- Take the requisite wire and cut it in the middle. Ensure that you find the perfect central line.
- Each piece should connect to an end insulator.
- Twist the piece back on itself for conducting an initial check.
- The other end of each piece should connect with the center insulator.
- Complete the attachment in a similar manner as you did for the end insulator.
- Bring the feed line and the center insulator together.
- Solder each connection.
- Hoist the dipole in the air with the aid of a few ropes.
- Check your dipole for readiness by doing a few short and low-power transmissions.
- This will help you measure the SWR. It should be less than 1.5 to 1, on whichever frequencies you opt to use.
- In case you find that the SWR is quite low at an extremely high frequency, or is the lowest at the higher end of the baud, it indicates excessive tightness at the end insulators. In such a case, loosen the connections.
- At the same time, increase the length of the antenna at each end. A few inches at either end should suffice.
- In case the frequency of the lower SWR is excessively low, you may decrease the length of the antenna again by a few inches at each end.
- Once you have adjusted the length of the antenna to the satisfaction of the SWR, it is time to wrap up! Securely wrap the wires at the end insulators.
- If there are excess wires at each end, trim them. Your dipole is ready!

You may go through the same steps if you wish to construct other simple wire antennas. The only thing that you may have to vary is the lengths. As for your dipole, just link it directly to your transmitter. Use coaxial cables as connectors. While using the dipole, ensure that the band is suitable for a 1/4-wavelength or any odd number of 1/2-wavelengths. To illustrate, if you have a dipole that is 66 feet in length, and works on a band of seven MHz, it will work equally well on a band of 21 MHz.

There are other types of simple and commonly used wire antennas too.

Full wavelength loop – Take the feed line and connect it to the center of the loop's bottom. Now, erect the loop, such that it becomes vertical. As a result, the feed line will find it easy to work broadside to the plane of the antenna's loop. After you construct it, you will realize that this is a larger antenna than the dipole. Additionally, it radiates slightly more signals in certain directions.

Random-length wire – In this case, all you have to do is to connect an open-wire feed line, measuring 15 to 35 feet at one end, prior to extending it as high and far as you can. In fact, you may even opt for a couple of bends. Use an antenna tuner, or balun, for converting the feed line to coaxial cable. Note that you will not be able to predict the behavior or performance of your random-length wire! However, you may use it as a temporary antenna, or even as a back-up one.

Inverted-V – You should ensure this dipole has adequate support at its midpoint. You should see angling of 45 degrees at either end. It helps that you need to provide a lone support for this antenna. It returns your goodness by working well in almost all directions!

Trap dipole – In this case, certain properly placed components come into play for isolating portions of the antenna. The isolations are in alignment with diverse frequencies. This enables the dipole to duplicate a simple 1/2-wavelength dipole, which works on a couple or more bands.

Multi-band dipole – Here, the wires feed at the center via ladder-line or open-wire feed lines. The whole construction is akin to an antenna tuner for covering many bands. Known as a doublet, because it refuses to go up to, or even beyond, a 1/2-wavelength on any band, the dipole is different from the actual 1/2-wavelength dipole.

To conclude, even if you do not possess the perfect backyard for setting up an antenna that matches your dream, go ahead with your experimentation. Use an antenna tuner for completing your experiments. You may bend or arrange wires as you wish, even at strange angles. Something is bound to work, eventually!

Vertical Antenna

It vies with the wire antenna on the popularity charts! Two of the commonest designs are the 1/4-wavelength and the 1/2-wavelength models. It helps that the vertical antenna maintains a low visual profile and does not expect a tall support. Furthermore, it is very easy to transport or move. This antenna has another name, omnidirectional antenna. This is because it is ready to radiate in all horizontal directions.

If you see the 1/4-wavelength vertical antenna, you may be surprised to note its close resemblance with the 1/2-wavelength dipole. Remember that this dipole displays a cut in the center and has turns at its ends. With regard to the vertical antenna, an electrical mirror of sorts makes up for the missing part of the dipole. This mirror is the ground plane or ground screen. When you see a ground screen, you will notice that it has a dozen or more wires that radiate from the base of the antenna. These wires lie on top of the ground. A feed line connects to these radials, as well as to the vertical tube/wire. Even the radials have connections to one another. Thus, a 1/4-wavelength vertical antenna should be easy to construct. Similar to the dipole, it should work well on odd multiples of the lowest design frequency.

If you opt for a 1/2-wavelength vertical, you will find that it measures twice the length of its 1/4-wavelength counterpart. However, unlike the latter, the former does not need a ground screen. This means that you may mount this particular vertical antenna on a mast or a structure which is at a height from the ground. This means the antenna is independent of the ground. You will need to connect the feed line to the vertical antenna. However, this feed line needs a special impedance matching circuit in order to work well with low-impedance coaxial feed lines. If your ham radio station has limited space, or you wish to acquire traditional antennas, you may enjoy working with ground-independent vertical antennas. You will be happy to know that both, the 1/2-wavelength and the 1/4-wavelength vertical antennas are capable of working on diverse bands. This is possible because of the use of similar techniques that come into play with wire antennas. If you wish to obtain multi-band vertical antennas for commercial purposes, they are available too. You may use as many as nine amongst the HF bands.

UHF/VHF Antenna

Generally, fixed stations opt for beam or vertical antennas, specifically if the band is beyond 50 MHz. The beam antennas come into play during the usage of UHF/VHF DXing on SSB and CW. This is for DXing. CW and SSB operators prefer to use horizontal polarization. This is because several of the UHF and VHF propagation mechanisms are more compatible with horizontally polarized waves.

Vertical antennas, on the other hand, prove exclusively useful for FM operations. In fact, these vertically polarized antennas are highly popular amongst the FM operator crowd, since the mode had its start in the mobile radio industry. It is very simple to mount antennas on vehicles in a vertical manner. Even back then, they opted for vertical base antennas in order to avoid cross polarization. This remains a universal belief even today. In case the beam antenna accesses a repeater at a distant place, the operator ensures that he/she positions the mount vertically.

In case you want to operate an all-mode radio, which you use for SSB/CW and FM modes, you will need to set up both horizontally and vertically polarized antennas. The most popular and affordable vertical antenna is the quarter-wave whip or ground plane one. Its functioning is the same as the antennas used for high frequencies. Keeping this in mind, several ham radio operators construct a two-meter ground plane, which is 19 inches in length as their initial project. If you want to emulate them, ensure that you extend the vertical antenna to 5/8-wavelength versions. This will help you receive slightly enhanced signals. In fact, if you opt for a two-meter vertical antenna, measuring 5/8-wavelength, it will double up as a six-meter vertical antenna with 1/4-wavelength too! You will not need to make any modifications!

UHF/VHF DXers use Yagi antennas too, at times. They like to have a reduction in antenna sizes at higher frequencies. As a result, these operators may opt to have several more elements than at higher frequencies. Do not be surprised if you around five to seven element beams on 50 MHz. Sometimes you may see over 20 elements at very high frequencies! However, these antennas, which provide high gains, possess narrow bandwidth too. In actuality, the bandwidth is too narrow for any kind of casual operations. Therefore, if you select a beam antenna on these bands, stop with three to five elements only. Let the antenna measure six meters in length. For higher frequencies, go for anything between five and 12 elements. You should find these small antennas easy to mount, as well as turn, if you use heavy-duty television antenna hardware.

While radiating, antennas tend to receive electromagnetic fields. These fields are a combination of electrical and magnetic fields. The conducting surfaces of an antenna are comprised of electrons. These electrons are in constant motion, moving back and forth. As a result, they create currents. These currents take the same direction as the oscillations in the electric field. The result is the flow of currents within the antenna. The flowing currents may cause the antenna to radiate a signal if the feed line becomes responsible for applying a field from a transmitter. Alternatively, they may cause the antenna to receive a signal, if the field is undergoing polarization.

When the radio wave's electric field aligns with the earth's surface, the resulting phenomenon is polarization. Since the movement of electrons and the electric field are parallel in alignment, the polarization of an antenna matches the direction in which you have arranged its conducting surfaces. To illustrate, vertical antennas undergo vertical polarization, whereas the majority of Yagi antennas undergo horizontal polarization.

Do you want the best communications? Then you must fix the antennas in such a manner that there is alignment amongst the electromagnetic fields created by each antenna. In case of non-alignment, the electric field created around the transmitter refuses to allow the electrons from the receiving antenna to move. This prevents current from flowing into the feed line. This indicates that the antennas have undergone cross polarization. The result is a substantial loss of signals. To illustrate, when you use Polaroid sunglasses, you have taken advantage of the fact that the majority of reflected light (glare) is horizontally polarized. The lenses are comprised of vertically polarized plastic, thereby eliminating glare through cross polarization.

Where is polarization most important? The answer is that it acquires greater importance at VHF and UHF. This is because these signals make an entry, with their polarization remaining intact to a major extent. At such times, on HF bands, the ionosphere goes into action. It scrambles the polarization to such an extent that the relative polarization loses its importance. Regardless, polarization has a significant impact on signal strength. You will know when you see the fading of deep signals, thanks to the occurrence of cross polarization. In fact, it is a common occurrence at high frequencies.

Portable and Mobile Antennas

When you encounter VHFs while using FM mode, it's good to opt for vertical antennas the majority of the time. This is because the wavelength at VHF is more compatible with a full-sized antenna. In contrast, mobile high-frequency antennas prefer to be smaller. In fact, although they appear reduced in size, they are similar versions of the vertical antennas on display at fixed stations. It is rare to find horizontally polarized mobile antennas for high frequencies. This is because their large size and mechanical considerations do not prove suitable for the same. Mobile high frequencies need to access maximum efficiency from short antennas, which is a challenge!

With regard to mounting mobile antennas, you may opt for permanent or detachable fixtures. For instance, the mag-mount models are the easiest to remove. These models have magnetic bases, which help them to attach themselves to metal surfaces quite easily. You may use these antennas from high frequencies through UHF. However, you may not enjoy looking at the installation, for it definitely is not as neat and clean in appearance as a permanently mounted antenna. Instead, you might opt for a trunk-mounted antenna when you require it for UHF and VHF. These models are semi-permanent in nature, and look much better than temporary structures. However, permanent mounts make for the best display of all! You could easily drill a hole in your car and set up a permanent mount for your mobile antenna. Whatever your choice, you may rest assured that all three types are almost equal in performance. Just ensure that you remove the antenna from your permanent, semi-permanent or temporary mount when you exit your car. Otherwise, you may encounter a theft or suffer during clearance (as in a car wash). It is possible to route your antenna cables under the car's seats, trim or carpet.

Where high frequencies are concerned, several mobile stations prefer the hamstick kind of antenna. These antennas remain attached to a mag-mount. Every antenna is comprised of a wire, which remains coiled on fiberglass tubes around four feet in length. They exhibit a stainless-steel whip at the top. This kind of an antenna functions on a single band. Since it is inexpensive, you may opt for an entire set to use for your car. You will have to change the antenna while changing bands.

A third design consists of resonators coming into play. These resonators attach themselves to a permanent base, such that they may function on diverse bands. If you wish to, opt for an assortment of multiple resonators. This way, you will be able to use many bands, without having to change the antenna every time you switch. You may mount the resonators and the accompanying fiberglass whip antennas on standard 3/8-24 threaded mounts.

In recent years, an adjustable design has made its mark in the marketplace. It possesses a top section that moves. This permits the antenna to tune over almost any type of HF frequency. This is the screwdriver antenna, which functions on a DC motor. You may see this kind of a motor in an electric screwdriver, which runs on a battery. However, this kind of antenna is highly expensive. The hamstick type of antenna is the least expensive. To conclude, any mobile antenna's performance is highly dependent upon the way it is mounted and installed. In fact, the variation in performance may be quite drastic to witness! Therefore, there is no way of guaranteeing good results anywhere.

With regard to UHF and VHF, if you bring portable antennas into the picture, the light and small products are very easy to pack and carry. With regard to HF, however, it becomes rather difficult to tackle the larger antennas. In case you are able to find some way to support a lightweight wire antenna off the ground, go ahead. You cannot do that with vertical antennas, for they need sturdy bases, as well as a ground system in place. Another option is telescoping antennas. You may opt for a few radial wires, along with mobile hamstick-style whips.

When an antenna does not exhibit a low SWR, it indicates that it wants a tuner. A tuner is useful for the transmitter, which will release full power. The outcome is the addition of expense and weight. Even smaller coaxial feed lines (RG-58 types, for instance) experience high losses. Therefore, before you rush off on a road adventure with your mobile antenna and feed line accompanying you, test their performances beforehand. You do not want to experience unpleasant situations on the backroads!

Feed Lines and Connectors

At the outset, do not be under the impression that choosing a feed line is child's play! It is not! Remember that the functioning of the feed line can have a positive or negative effect on the entry and exit of your signals. Therefore, you need to select a feed line that will award you the least loss possible. In case you decide on an open-wire feed line, you will incur extra expenses on the addition of an impedance transformer or tuner. This is for awarding a 50-ohm load to your transmitter. In this scenario, you win some and lose some! Therefore, it would be best to purchase feed lines in spools measuring 500 feet from an authentic distributor. This will help you save plenty of money! In case the entire spool proves to be highly expensive for just one person, you may request another ham radio operator share the material and its costs. You may split with more people too. This will prevent you having to make do with 90 to 100 feet of feed line every time you make a new purchase. You may do the same when purchasing coaxial conductors.

Has someone told you that you can buy used cables and save on money? Well, it is indeed an option, but you have to ensure that the seller is someone authentic. It would be tragic if you were to receive old cables which were lossy simply because water had entered their ends. Alternatively, water might have entered splits or cracks in the cables' jackets. Sometimes you may see a cable that has acquired a sharp bend. Since it has been in that position for a long time, you will find it very difficult to unbend it. Additionally, it becomes possible for the central conductor to penetrate the insulation. This will cause the development of changes in the cable's properties, or even a short. You will witness this aspect of migration specifically in cables with foam insulation. Therefore, when you opt for used cables, spend some time examining them carefully. Purchase them only if you are sure that they will work for you, without causing any headaches!

To check for second-hand cables:

Do a thorough examination of the cable. You should be able to see a shiny and smooth jacket. This means that there should be no cracks, nicks, scrapes or dents on the cable. Similarly, check if there are deposits of adhesive material or tar on it. The deposits are the outcome of the cable resting on the outside of a building or on a roof. It could be that the owner even secured the cable with something, to ensure that it did not move back and forth. When you are through, make a slit in the jacket at either end. Let the slits measure one foot each.

Now, inspect the braid. What is it like? It should exhibit a smooth shine. There should be no signs of discoloration or corrosion on it of any kind. When your investigation of the braid is complete, slip it back into its original place.

With regard to the center insulator, if it is solid in nature, it should appear very clear and clean. In case it is made of Teflon synthetic or foam, the insulator must appear white. Sometimes a cable displays a connector at the end. If so, you will find it difficult to research the condition of the cable unless you cut the connector off. It could be that the original owner of the cable had the connector newly installed. Then you will have to ask the dealer if you may cut off the connector, while providing the assurance that you will replace it with another one. Are you able to cut the connector? If the answer is in the negative, then do not opt to purchase the cable.

In general, the operators of ham radio stations go for BNC, N-type and UHF connectors. They are all standard RF connectors.

BNC

BNC stands for Bayonet Neill-Concelman. Paul Neill was responsible for the creation of this connector, way back in the 1940s. He wanted it specifically for the military, which preferred smaller radio applications. Neill hoped that this connector would prove useful in accommodating such applications. The best part of the BNC is its bayonet. With just a quarter turn of this connection, it becomes possible to create a positive lock.

You may buy this connector in two models—50 Ω and 75 Ω. It is imperative that you recognize which connector you are working with, as they appear very similar to one another at first glance. These models even match one another mechanically. The ability for carrying signals is dependent upon your application. BNC connectors come into play for low power, going up to 100 watts, running on frequencies through 440 MHz. Ham radio users tend to keep their usage under 4 GHz. This ensures that there are no issues. However, if the frequency goes beyond 4 GHz, there will be problems because of the connector's bayonet-style sheathing. The BNC is losing its popularity in modern times. This is because of the entry of the SMA (discussed below), impressing upon people that the BNC is large and not as mechanically sound as a connector should be.

SMA

This is the Subminiature Version A. You are sure to come across this as the first choice in connectors in the world of ham radios when you begin your career in amateur operating. The lightweight connector is all-purpose in nature and specifically useful for small applications. An example is the handheld radio. The bonus is that the SMA is capable of carrying signals which range from 0-18 GHz. This is outside the range of a majority of ham operations. Then again, it is one of the securest of connectors, thanks to its a threaded connection. Probably, its inclination towards perfection lies in the secret that the inventor developed it for the military. The military always expects something very strong and powerful. Therefore, the SMA, despite its small size, receives praise for its strength.

This connector came into the public eye in the 1960s, with the aim of offering a simple connection interface to the RF coaxial cable. Probably, this was a competitor to the creation of the 75 Ω'F connector, which became popular in the early 1950s. The 75 Ω'F connector was a common feature for television applications. Therefore, if you should set out to obtain an SMA, make sure that you do not become confused between the two! The SMA and F connectors look very similar. True, their dimensions vary. However, the major difference lies in the impedance. It is different for both; wherein an amateur radio station requires 50 Ω and the F connector provides 75 Ω. If you try to connect the two, it will only lead to disaster! As mentioned earlier, modern handheld or HF radios prefer the smaller SMA connector, in comparison to other connectors. Furthermore, the threading is so unique in design that cross-threading is not possible. Manufacturers only expect the SMA to grow in popularity in future!

Female SMA

The name sounds strange for a connector, doesn't it? It is the output of a radio-manufacturing phenomenon in China, wherein brands like Yanton, Baofeng and Leixen are involved. However, it possesses exactly the same kind of interface that the SMA connector does. It is wonderfully compatible with specific types of handheld radios. However, it is also incompatible with the majority of standard antennas, and there are no adapters. The term female comes into play here in order to indicate the switching that takes place between the male and female posts on radios and antennas. If you observe the standard SMA connector, the male post is on the antenna. It is the reverse with the SMA female connector, which uses all the same parts, unlike the RP (reverse polarity)-SMA that reverses the external housing. It is possible to bring together a female SMA antenna and a standard SMA antenna for threading to one another. It helps that the major manufacturers of antennas for ham radio stations ensure that they have female SMA connectors in stock.

TNC

The abbreviation refers to Threaded Neill-Concelman, since Paul Neill also invented this connector. It is a duplicate of the BNC, except for one detail. The TNC is a threaded connector which offers greater mechanical security. The threading enables the propagation of signals easily. These signals are of a much higher frequency than those that a BNC connector can carry. The TNC connector can go up to 11 GHz, with an impedance of 50 Ω. Regardless, this connector is not very popular, and ham radio operators do not seem to have it much in evidence at their respective stations.

UHF

This connector can go up to two meters, with the ability to tackle complete legal power. You should be able to see this kind of a connector on a majority of mobile radios, as well as at high-frequency base rigs. The UHF connector is probably the oldest amongst all connectors, for it has been in existence even before the advent of World War II. Ham radio operators regard it as an old-school connector! In those days, people considered anything up to 30 MHz as UHF. As a result, the entire connector system became the UHF connector. An alternate name for it was PL-259 connection, wherein the PL-259 plug became the male component of the system and SO-239 socket became the female component.

The UHF connector has been under observance for quite some time now, thanks to the fact that there have been significant losses in performance as soon as radio frequencies stretch beyond 50 MHz. Experts have discovered this through intensive testing. As a result, the UHF is losing its popularity.

N-Type

This is a waterproof connector which can progress through 1,200 MHz. If you install it properly and maintain it, it can yield full legal power. The N-Type connector is the brainchild of Paul Neill, who used to work at Bell Labs. It carries his name. The N-Type has found favor as the most versatile connector in the world of ham radio operators. It is also one of the most mechanically sound connectors amongst all the other connectors in use today. Once again, this connector was a gift for the military, way back in the 1940s. Over the years, the connector has only grown in popularity. It is capable of handling up to 11 GHz. The most important feature of this connector is its waterproofing. There is a rubber gasket present in the seat of the male casing, which is responsible for the same.

Repeater installs, or base station setups, go for this connector in a huge way, due to its waterproofing feature and soundness in mechanical functioning. Nowadays, ham radio operators have joined the fray too, even going to the extent of replacing UHF connectors that had come into play on their mobile radios. They opt for aftermarket Type-N connectors. Thus, the N-type is rapidly acquiring the label of best of the bunch!

Now that you have understood the detailed aspects of each connector, you should understand that good-quality materials are affordable too. Additionally, you should always obtain the best for your ham radio station, if you wish to be a successful operator. Therefore, do not scrimp on the purchase of high-quality connectors. You do not want to witness water seepages, corrosion or physical breaks due to your inexpensive connector working loose! These occurrences will only eat away your valuable signals. If you are confused, just buy the PL-259, a plug that is compatible with the ends of coaxial cables. You should be able to save on money if you purchase in bulk. Note that you should stay away from shiny nickel-plated connectors. You will find it difficult to solder them.

Another question that may be bothering you has a connection with the aspect of crimping. You may wonder whether you should, or should not, crimp. It is easy to install a crimp-on connector, for obtaining reliable service indoors. An environment of low humidity and low temperature will support its functioning. It is easy enough to obtain crimpers or crimping tools for affordable prices. In case you require several connections, it might be good to get a crimp-on connector. At the same time, it is possible to opt for a silver-plated connector with appropriate soldering. It outperforms a crimped connector every time. You may even use it outside, unlike the crimp-on connector, which is only useful indoors.

D. Support for the Antenna

Antennas are available in all manner of sizes and shapes. You may find something the size of a mere finger or something so large that it's hundreds of pounds in weight! The only common factor amongst all antennas is that they work well only if there are no obstacles around. It is imperative to ensure that your antenna 'breathes' freely and remains away from ground level. After all, the majority of obstacles tend to reside at the ground level only. Keeping these considerations in mind, peruse safety information linked to the functioning of antennas and their support systems, prior to working with them. In fact, you may even consult an arborist, if you have to work with trees around the place and need advice about raising towers and masts in the area. You should have some information, as well as common sense, about climbing too.

Masts and Tripods

A mast, whether you have had it constructed from metal or wood, is affordable. In fact, it is one of the best ways to provide support to your antenna. You may maintain a distance of up to 30 feet between the ground and your antenna via this method. Bearing this in mind, ask yourself if you can build your own home-brew mast or not. If you are handy with tools, go ahead. You need not worry about instructions or guidance, for there are numerous articles available in all kinds of ham magazines. Such a mast is excellent for holding up wire HF antennas, as well as UHF/VHF antennas. The UHF/VHF antennas include small beams and verticals. At the same time, if you merely need a support for your UHF or VHF vertical antenna, you will not have to get something heavy at all. As an alternative, you may create a self-supporting mast which does not need guy wires at all. Before you make this kind of a decision, however, check out the changing weather conditions in your area. Is it subject to high winds? Then you cannot afford not to buy guy wires. It is the same if you fear that your mast may fall victim to a side load, such as you may see in a wire antenna. You will need to use guy wires.

There is an option that is available in the commercial arena—a telescoping, push-up mast. This kind of a mast has a role to play in holding up the small antennas attached to television sets. You may have seen them installed on rooftops, quite often. It is possible to find a push-up mast measuring up to 40 feet too, with guying points attached to it. Alternatively, you may build a mast using sections of metal television antenna mast. You just have to obtain them in short sections and stack them. With regard to guying points, however, you will have to create your own. Then again, it is not possible to climb up sectional or telescoping masts. Therefore, you will have to work out your own strategies for mounting the antenna, as well as erecting the entire assembly. A way out is to mount a couple of sections of stacking mast onto a chimney. The chimney will provide adequate support to a small vertical antenna. Of course, the easiest thing to do is just walk into a hardware store, or even a home improvement store, which supplies push-up and television masts, and purchase what you want! Such stores even supply the accompanying paraphernalia required for mounting and guying.

If you would like to go even beyond the mast, opt for a roof-mount tripod. Light tripods suffice to hold small amateur antennas, as well as television set antennas. The larger tripods tackle mid-sized high-frequency beams. If you live in an urban area, a tripod may prove to be the best solution for you. It is the same if you reside in a subdivision, where you may fail to receive permission for a ground-mounted tower. To suit your specifications, you may purchase a tripod from a manufacturer of antennas and towers.

Towers

The sturdiest support for an antenna is a tower. A tower may go up to 30 feet and beyond. There are diverse types of towers available in the ham marketplace. You may opt for a self-supporting or unguyed structure, a tilt-over tower, a guyed construction or a multi-section crank-up. Even the largest of antennas or the highest of heights finds it difficult to defeat a tower! However, you must keep in mind that the construction requires substantial effort and space. There are manufacturers who will provide you with everything you need to set up your tower. However, you will need legal permission to erect a tower.

Self-Supporting Tower

This is a triangular construction, albeit in cross section. The raw materials are comprised of truss-like sections, similar to lattice towers. Designed from angle, solid rod or pipe, the self-supporting tower exhibits lattice-type constructions, each possessing three or four sides. The base has to be large and made of concrete. This is to enable control at the center of gravity. However, the self-supporting tower is definitely less costly in comparison to a multi-section crank-up tower. Similar to the lattice tower, this structure has good carrying capacity too. The tower may extend up to 100 feet. The snag is that you may find it difficult to mount an antenna along the length of the self-supporting tower. It is easier with a lattice tower, which has supporting legs that progress vertically.

A self-supporting tower works well for almost all wireless communication applications. You may find it an ideal choice if your space is limited. You do not have to spend much on transporting it or installing it. Its versatility is evident by the fact that it comes into play as a cellular tower, wind tower, broadcast tower, homeland security tower, radio tower and internet tower.

Tilt-Over Tower

This kind of tower is comprised of a lattice construction with hinges in the middle of it. The hinges permit the top section of the tower to pivot towards the ground with the aid of a winch. The mechanical considerations of such a tower demand that it may not be above or below 100 feet in height.

Crank-up Tower

You will see lattice sections or telescoping tubing coming into play for a crank-up tower. It is possible to use a motorized or hand-operated winch for raising and lowering the tower. The winch functions on a cable and pulley arrangement. In general, when the completely nested crank-up tower goes up, it's 20 to 25 feet high. As a result, there is less visual effect on the neighborhood. Additionally, you will be able to climb up the tower and work on your antennas whenever needed. Something else that you will see in such a tower is a base that is capable of tilting. This unique construction helps in the transportation of the tower, as well as its erection. It is necessary to give the crank-up tower a solid and humongous concrete foundation. In fact, the foundation spreads over several cubic yards, in order to ensure that there is no disturbance to the center of gravity. It also ensures that the center of gravity remains below ground level, in order to prevent the tipping over of the tower when it extends fully. Such precautions are necessary for a crank-up tower possessing no guy wires. You may use such a tower if you reside in a small area where guying is impossible. You may have the tower set up to a maximum height of 90 feet.

Welded Lattice Tower

This is one of the commonest options amongst ham radio operators. Constructed from 10-foot sections of steel tubing, it possesses welded braces too. It is imperative that you guy it. Alternatively, you may fasten it to some kind of supporting structure, such as a house. The house will provide support up to a height of 30 feet, or even more. Then again, the tower requires a footing. This footing should take the form of a concrete base, measuring several cubic feet. As an amateur ham radio operator, you would do well to have a lattice tower comprised of 12-24 inches on a side. You should be able to use these measurements to set up a tower going well beyond 100 feet in height. You need not worry about the strength of your tower. It is strong enough to support quite a few large high- frequency beam antennas, if you ensure that the guying is appropriate.

Now, similar to all other things that you purchase for your ham radio station, you may opt for a brand-new tower or a used one. It is the same with the mast. However, you will have to be careful. If the tower or the mast has not been stored appropriately, it may have suffered exposure to diverse elements in the environment. As a result, you may witness weakening in the welds and supporting member, or corrosion. Or someone may have disassembled the mast or tower inappropriately, resulting in damage, albeit in subtle and unnoticeable ways. In fact, you may not even be able to detect the damage, especially in the separate tower sections. If you find that the tower or mast exhibits cracks or warping, you may rest assured that it has fallen sometime or other. It will not provide safety for your antenna at all. Thus, it would be best to ask an experienced expert to accompany you whenever you go shopping for supportive structures for your radio's antenna.

You will be happy to know that you may build your own self-supporting tower, if you have the inclination to do so. It is easy, because you may use all kinds of unorthodox materials for the construction. For example, you may use well casing, telephone poles, light standards, etc. If these innovative materials are capable of holding up your antenna, it is because someone from the ham radio community has tried them in the past and shared the knowledge with others! The challenges lie in transporting this mast/tower. Even erecting it is not easy. Furthermore, you should find it safe to climb, after its complete construction. You will have to climb it to create a sturdy and strong antenna mounting structure at the top of the tower/mast. Despite everything, if you still need help, just peruse the online forums. There are many experienced ham radio operators who are more than willing to share their knowledge and expertise. The topics under discussion include locating True North, mounting verticals on a rooftop and so on. Several experts enter these forums too, and one-to-one discussions are possible.

Trees

Amazingly, a tree can prove to be the best support for your antenna! In fact, it is the oldest kind of support, in existence long before mechanical constructions appeared on the scene! The handy tree is capable of holding up wire antennas best, since these antennas acquire horizontal positions. These antennas may go in for horizontal support ropes too. You will rarely find rotatable masts or antennas installed on trees. Even the tallest and straightest of conifers will not do for this job! This is because no tree can handle the mechanical complexities of such structures. There is the possibility of damage to the tree, as well as the occurrence of mechanical interference between the tree and the antenna. However, if you desire aesthetic appeal and you possess the right type of antenna, you may consider a sturdy tree for your ham radio setup. After all, the tree does not charge you for its services!

To acquire support from the tree for your antenna:

First, obtain a pulley or halyard. Next, push it up into the tree at the maximum height possible. This should not be hard to attain, if you are confident of your climbing skills. Alternatively, you may request someone else climb the tree on your behalf. This will enable you to position the pulley in place, manually. Otherwise, you will have to figure out a strategy for sending a line up, through the branches of the tree, such that you can haul up the pulley easily. See if you can just throw the support line of the antenna over a branch, easily. At the same time, be aware that when the line is touching the bark of the tree all the time, it is bound to suffer from rubbing and chaffing. This can cause the line to break. The situation is worse if there is a storm, especially at night. You may find it frustrating to discover that your communication link has gone haywire, just prior to establishing an important contact! Therefore, think carefully before you act.

In case the line manages to stay in place and is safe, the tree may create problems. It may strive to grow around the line. As a result, you will find the raising or lowering of your antenna next to impossible. Therefore, if you want to use a tree to support your radio's antenna on a permanent basis, you might do well to consult a professional who knows all about trees and can undertake the job in an expert manner. This individual will have adequately rated and sturdy materials on hand. Even if the fees seem high, the effort will be well worth the price. Alternatively, you may contact an arborist and request guidance.

All About the Rotator

Do not confuse this with the rotor. A rotor is a component of a helicopter. It has nothing to do with ham radio stations at all! What you need to know about is the rotator. This refers to a motorized gadget which positions itself on a mast or tower. Its job is to point the antenna in diverse directions.

In case you decide to opt for a rotatable antenna, you will have to check out the rating. The rating of the rotator takes place in connection with the wind load. You may check the wind load by observing the square feet of antenna surface. You may even contact the manufacturer of various antennas to get the ratings for wind load. This will help you to find out what the wind load of each rotatable antenna is, as well as the size of rotator that you require. Thus far, Hy-Gain has been the most popular choice for rotators, but you might check out other brands too.

The rotator has a special function. It has to turn the antenna in various directions. At the same time, it must also ensure that the antenna does not succumb to wind pressure. After all, the wind strives to turn the antenna on which the rotator is set against the braking mechanism of the rotator. It is imperative that the rotator use all its strength to keep firm control over the antenna while the braking mechanism is in the off position. When the brake is off, the antenna is free to turn against the wind. Since the rotator must perform the dual functions of holding the antenna, as well as turning it, it needs to exert torque on the foundation that is supporting the antenna. This foundation is the mast.

As long as the rotator is capable of exerting greater torque than the wind's pressure on the antenna, everything will remain normal. It is not easy to calculate or measure how much torque the wind is applying on the antenna. Therefore, it is necessary to assume that the wind is moving at maximum speed. In contrast, it is easy to specify the area of the antenna. Keeping this in mind, we relate the wind's torque directly to the antenna area. This is why the rating of a rotator is equivalent to the square feet of wind load. In case there are multiple antennas linked to a lone rotator, you will have to bring all the individual loads together and add them, for a total rating. Since you are an amateur operator, you may obtain a rotator with an initial rating of three-square feet.

Next, how would you like to set up your rotator? Would you like to set up it up on your mast or tower? Sometimes an adapter needs to enter the picture too, since the structure demands it. In general, the antenna goes onto a pipe mast, which stays in place via a mast clamp resting on top of the rotator. Whenever you mount the rotator on a mast, you must ensure that the antenna stays at the top of the rotator. This is to reduce side loading to the minimum. In case you decide to build a tower, you must ensure that the rotator remains attached to a rotator plate. The rotator plate refers to a shelf lying inside the tower. This shelf enables the rotator to rest upon it. The mast moves through a thrust bearing or sleeve. Thrust bearing refers to a collar or tube, which holds the mast in the center, just above the rotator. When you use a bearing in this manner, you help in preventing the side loading of the rotator. Therefore, you may mount the antenna much beyond the tower top, provided the mast is wonderfully strong.

You may control the rotator with the aid of control box. This control box refers to an indicator assembly, which you must install in your shack. With regard to connections with the rotator, you may use a multi-conductor cable. Finally, when you install the feed lines to connect to your antennas, ensure that the installation permits accommodation for the rotation of the mast and antennas.

You may feel that there is nothing wrong in using a second-hand rotator, especially if it will help you save money. However, this may prove to be a risky purchase. It's all right if you buy it from a trusted source and can use it successfully as your first installation for some time, at least. You must remember that you install rotatable antennas in exposed locations. They may wear out sometime or the other, albeit in ways that may not be immediately visible. As a result, the rotator may create the impression that everything is fine, when it is actually not. True, it turns well on the bench. However, when heavy loads enter the scene, it may stall, slip or jam. You will then have to replace the old rotator with a new one, leading to further expense. Therefore, save yourself all this trouble and just buy a new rotator at the beginning itself.

E. Accessories for Your Radio

You have the freedom to test all kinds of gadgets at your leisure, in order to enhance the functioning of your ham radio station, or even to make it special in the ham radio world. However, there are some accessories that every operator of ham radio needs.

Microphones

The majority of ham radios possess hand microphones, provided you have purchased new ones. In case you buy a second-hand radio, you may not find the hand mic at all, since it may have disappeared long ago! Even if you do find it, it may prove too worn out to use effectively. Regardless, the hand mics that diverse manufacturers supply tend to be quite good. They should suffice to help you begin your operations. Even if you do not find the microphone completely satisfactory during the initial stages, you may always upgrade it later.

Now, if you are a ragchewer, you may be glad to discover that some microphones possess the feature of audio fidelity. This audio fidelity comes with a wide frequency response. You may have noticed how people participating in contests, as well as net operators, love to have their hands free. Towards this end, they use headsets with attached boom mikes. They may bring these boom mikes in front of their mouths, in order to speak into them. You can opt for the same. With regard to handheld radios, however, it is more convenient to get a combined speaker-microphone accessory. You may attach the mic to your shirt collar or pocket, while plugging it into the radio at the same time. There is a wire that goes from the mic to the clips on your shirt/pocket. In case you are lucky, the manufacturer of your brand of radio may decide to offer a premium microphone too. This offer could come in the form of an accessory or an option. Whatever the case, just ensure that you obtain a top-quality microphone and headset.

It all depends upon how your microphone responds to frequencies, in order to check the on-the-air quality of your ham operations. These responses make a vital difference to your operations. For instance, if you manage your activities in crowded conditions, the microphone's audio will probably cut through the surrounding noise. However, this will happen only if the microphone's response places emphasis on the mid-range, as well as higher frequencies. Sometimes a microphone reveals a selectable frequency response. For example, if you are engaged in casual contacting, your voice will sound very natural to the other person's ears. When it is time for some DXing, the response becomes brighter automatically. Thus, in all probability, you might experience some confusion in choosing the right microphone for your ham radio operations. What you can do, therefore, is to ask a close friend to conduct an over-the-air check with you. This person should be able to give you appropriate feedback, as he/she knows your voice. Alternatively, you may request that all the people you contact offer their opinions on how your voice sounds on the air.

Keys and Keyers

People, who are enamored of Morse code may access thousands of keys. In fact, this selection has spanned a century of history. Those who enter this field go for the straight key during the initial stages. Later on, they move to the electronic keyer and its accompanying paddle. You could do the same, for you are still an amateur operator. In case you feel that you will be using CW quite a lot, it is best to go in for the keyer and paddle route straightaway. After all, modern-day ham radio rigs include a keyer. In fact, this is the standard option. You should find it easy enough to plug the paddle into the radio and begin your operations.

Operators dealing with CW, feel that the choice of a paddle is very personal. Therefore, if you want to be like them, go in for a trial prior to purchasing anything. If you approach hamfests, the conventions of amateur radio enthusiasts, you should be able to garner a lot of valuable information. You should be able to contact at least a couple of key-bug-paddle collectors. They will guide you regarding diverse styles.

Is your final choice an external keyer? If yes, then you may select from finished models and different kits. It helps to have programmable memories. You will be able to store common information in them, such as a CQ message, your call sign, etc. You could even try out something as novel such as putting your keyer in beacon mode. This will forward a CQ message that you stored earlier. The repeated message may even reach a dead band. This means that even when everybody is listening, but no one is transmitting, the band seems to be dead. However, even the dead band may lead to some place which is even more exciting to explore! Many computer programs send codes from keyboards. You should find such software listed somewhere online.

Finally, there is the voice keyer. This refers to a device which can capture short voice messages. The device also plays these messages back into your radio. It will sound as if you are speaking! Some voice keyers are standalone units. Other keyers prefer to utilize sound cards. Whatever type of voice keyer you may opt for, they come in handy for calling CQ (code used by wireless operators), contesting, DXing and so on. Certain models, such as Super Combo Keyer and MJF Contest Keyer, store both voice messages and CW.

Antenna Tuners

New radios have antenna tuners inside them. However, this may not suffice to do the job in certain situations. At such a time, you will require an external unit as well. This is because internal tuners function within a limited range. This range is suitable for many antennas. However, you may not be utilizing your antennas in alignment with what their design offers. You may not even be using them at optimized frequency. As a result, your antennas may come up with impedance that your rig's tuner finds terribly difficult to handle. In case you are using balanced feed lines, you may have to use a tuner which is capable of handling the jump over from coaxial cable to open-wire feed lines.

Antenna tuners are of diverse sizes. They may be extremely tiny, QRP-sized units. They may be humongous in size, displaying boxes of full power, which may be larger than several large radios. Considering that there is an array of antenna tuners to select from, you would do well to purchase a specimen that you may use comfortably. In fact, its comfort level generally goes beyond the maximum power that you want to use during your operations. In fact, it would be best to go for something that is compatible with balanced feed lines. Other features that you may like to have include the capacity of the antenna tuner to switch between varied feed lines and an SWR meter.

Although they carry the label of antenna tuners, these gadgets refuse to tune any kind of antenna at all! They are actually impedance matchers. An impedance matcher is akin to an electrical gearbox, which helps to transform the impedance that occurs at the end of a feed line to 50 ohms. Radios like to see 50 ohms always in evidence. The antenna is ignorant of everything that is going on! Then again, having an antenna tuner is not enough. You require a dummy load too. This dummy load refers to a big resistor, which has the capacity to dissipate the total power of your transmitter.

With regard to HF transceivers, if there is an MFJ 260C in place to tackle 300 watts, they are happy! Some loads are of high power, such as Heathkit Cantenna or MFJ 250. They ensure the immersion of the resistor in cooling oil. Thanks to the presence of the dummy load, your transmitter will face least interference during the process of tune up. However, it may not be possible to use a dummy load for UHF/VHF. Prior to buying anything, you will need to peruse the specifications regarding frequency coverage.

Controlling the Radio

Every modern radio displays an RS-232 control interface. This is for enabling you to monitor and manage every manner of function undertaken by your radio. This facility has prompted several radio control programs to place front panels on computers. The popularity of these programs has led to some establishments offering a specific radio control package along with the radio. It is your choice to download this software, or purchase it separately. Several third-party programs have become an integral part of logging software. In case you want to know more about radio control interfaces, you might consider joining a group where users describe their experiences, share tips, etc.

The Digital Mode

The digital mode is extremely popular, even in ham radio operations. In this case, the ham radio operations that have mostly been taking place on digital modes are equally quickly converging on the sound card. In fact, the sound card has become the standard option for sending and receiving data. Thanks to the presence of an interface that combines the simple audio functions and transmissions, it is possible for your radio and computer to bond. They become an extremely powerful data terminal. If you plan to use PACTOR II and CLOVER, which are proprietary protocols, you will have to use specific software and hardware from the developers of these protocols.

Chapter 9—The Licensing Regulations and System

Ham radio is very different from other types of radios that are available to the public, and one of the major differences is licensing. You need to get a license for using this service. Therefore, understanding the basics around the ham licensing system is important so that you know how to get a license, which one to opt for and how much it should cost you. If you have an understanding of the process, you can easily deal with the way it works.

Amateur radio is one of the different types of radio services that utilize radio waves for communication purposes. Apart from amateur radio, radio waves are used by services such as aviation, public safety, broadcast and so on. And since multiple services are using this band, the FCC wants each of these stations to acquire the license to use it and adhere to their regulations in order to keep this license. And that's what the ham license is all about—it gives you the authority to transmit messages on the frequencies that you are licensed to use. So let's dive deeper and understand how the licensing system works for amateur radio services.

An Overview of the Amateur System

An amateur service is a licensed service that approves the application of every ham who wants to transmit in this band. Although the regulations are quite old-fashioned, licensing is a must because of the following reasons:

- Hams use the amateur service to communicate with people across the globe, without any intermediate system to keep an eye on their activities.
- Due to the scope of this radio service, hams require a minimal amount of technical and regulatory background so that they can efficiently work without disturbing other services, such as aviation and safety services.

By ensuring the high quality of licensees, the system that regulates and maintains the licenses ensure amateur services are used to the best of the knowledge of each user. And that's the only way to ensure these services are set apart from other unlicensed services, which are not authorized by the FCC. Ham radio services do all the good things in return of the space they have been given in the radio space.

All the radio services administered by the FCC come with their own set of rules and regulations, wherein each rule is defined separately and given a unique number, for instance—Part 97.101. Within each of these rules, there are individual parts that talk about the practices to be used. Although the numbering might look complicated in the beginning, it helps in finding the exact rule in no time. Part 97.1 is the most important part of the licensing as it mentions the basis and purpose of the amateur radio service. It talks about the mission and what the service intends to accomplish. According to the explanation given in this part:

> *The rules and regulations in this part are designed to provide an amateur radio service having a fundamental purpose as expressed in the following principles:*
>
> (a) *Recognition and enhancement of the value of the amateur service to the public as a voluntary noncommercial communication service, particularly with respect to providing emergency communications.*

The word that should be noted here is "noncommercial," which means hams are not supposed to get paid for their services (although there are some exceptions to this) and must work on voluntary basis. These voluntary services include performing and promoting business activities on air. The services are really important as they handle the emergency scenarios and respond during disasters to connect to the affected areas. Nothing can be more important than responding to emergencies, and therefore this is one of the reasons we have these amateur services today.

Hams are known for their history of inventing things. They are known for how they best utilized the so-called "worthless" shortwave bands after World War I for long-distance communications. And they haven't stopped. Even with all the inventions and discoveries happening around the globe, they are still inventing new systems to help the public. They not only play around with radios to come up with some interesting ideas, but they train to operate them to be used in various useful ways. There are several events and exercises that are conducted to ensure hams' communication skills are continually developed and upgraded. After all, having a bunch of experts who can handle ham radios is of great value to the entire world, be it the public or the military industry.

Ham radio service is like an international passport to friendship. There cannot be anything better than connecting with someone who is oceans apart, whether it is using Morse code on the high-frequency bands or an IRLP chat. Hams have their own unique ways of making contact with people in different parts of the world, with no intervening systems or organization. They just have to abide by the rules and regulations stated by the FCC. This license allows the hams to operate anywhere within the 50 states that are under the control of the US government. Hams can also operate from other countries that have reciprocal operating authority with the US.

With the FCC as the central body in charge of setting rules and regulations for ham radio service, here are certain definitions one should be aware of:

The amateur service is a radio communication service that is created for the purpose of intercommunication, self-training and investigations by authorized people who do not have any interest in making money out of the process. Here, the radio communication service is governed by a set of rules and regulations defined by the body that administers various activities such as marine, amateur, etc.

An amateur operator is someone who happens to be the control operator of an amateur station. According to the amateur service rules, if you have an amateur license, you cannot operate a station

An amateur station is a station that consists of the apparatus needed for carrying on radio communications. This means that any radio station that complies with the rules of amateur services can be an amateur station.

So, if one clearly understands these three terms—amateur service, operator and station, it is easy to understand the set rules as this is where it all starts from.

Types of Licenses

To get an amateur radio license, a person needs to not only pass the test but also fulfill an important criterion—he/she shouldn't represent any foreign government. However, there are no other requirements to obtain the license as ham radio offers equal opportunity to everyone.

An amateur radio license has two parts to it:

- An operator license
- A station license

While an operator license gives you the authority to operate an amateur station, as per the rules set by amateur service, the station license allows you to have an amateur station. when the two are combined, it is an amateur operator/station license. Not all the services are comprised of these two parts; certain services grant them separately—for example, broadcast stations grant the two licenses separately as the employees operate the equipment.

There are three types of amateur radio licenses:

- General
- Technician
- Amateur extra

Each of these classes comes with a different set of operating privileges and frequencies that expand from one level to another, along with the comprehensiveness of the exam. When you pass the advanced levels, you earn more privileges. In addition to all this, there are other license classes—

- The novice
- Technician plus
- Advanced

Technician class: Almost every ham starts with this class license and once he gets it, he is allowed to access all the 17 ham bands that occupy the frequencies of 50 MHz or higher. The privileges given to the technician licensee include all types of communication and operation at the maximum legal limit. The test to obtain a technician license is comprised of 35 multiple choice questions on different topics. To pass, a candidate needs to get at least 26 answers right.

Additional privileges are awarded to the candidate who passes a five-word-per-minute Morse code exam. This includes certain transmitting privileges in four of the traditional shortwave HF bands.

General class: Having passed another 35-multiple choice question exam and a five-word-per-minute Morse code exam, hams gain access to more frequencies and privileges. General-class licensees have access to almost the entire spectrum of frequencies with only certain part of HF bands excluded. Although the test to get this license includes the same topics as those in the technician class, they are more detailed. In addition to this, there are certain new topics that an experienced ham is expected to understand. So, once a ham passes the test for technician class, he can upgrade his knowledge to general class by taking this test. And once this is achieved, he achieves a great milestone. All of the important frequencies available in the HF bands are available to the general class hams.

Amateur extra class: Although we say general class can access almost everything, it is not *everything*. There are still certain frequencies and privileges that can be accessed only by amateur extra-class licensees. This includes the lower segments of various HF bands which are normally accessed by expert operators as it is their prime operating territory. Therefore, if the hams are interested in contacting rare foreign stations, or wish to access frequencies of their choice, they should take the Amateur extra class test to get this license to the top-level class. The exam for this class is comprised of 50 multiple choice questions and at least 37 out of these 50 must be answered correctly in order for a person to pass the test. Unlike the other two categories, this test covers certain additional rules and regulations around complicated operating and advanced technical topics. And all those who pass this test take pride in having attained the amateur extra class license.

Along with understanding of the different types of classes and the tests associated with these classes, there's a special topic—the call signs. A call sign is how you are identified on the air; it is your unique radio name. Each license comes along with a unique call sign but hams can change them once or twice before deciding on a final one. And sometimes, these call signs become your identifier even when you are not on the air; for instance, your last name gets replaced with the call sign. Call signs have a common structure; they are combinations of alphabets and numbers, for example LAØN, and each letter and number are pronounced individually.

Ham call signs are created from two parts—the prefix and the suffix. When they are combined, they make a unique sign which helps in identifying a ham. For example, there will be a lot of call signs comprised of LA and ØN, but there will exist only one call sign LAØN. The prefix can have one or two letters and a numeral, and the suffix consists of one to three letters. The importance of the numbers and alphabets used in the call sign is that they identify the country that issues the license and also mentions where you are in that country. For call signs used in the US, the numeral describes the call district of the license.

Each country is assigned a block of character groups for the prefix that helps the government to assign the licenses to different radio services. The licenses given in the US consist of letters A, N, K or W in their call signs. These letters are used by not just hams, but even broadcast services have call signs such as WKL, etc. Canadian call signs use VE as the prefix. Germans have it as D, while the Brits use M or G.

The call sign also tells the class license you have. And as you upgrade your class, your call sign changes to reflect the class license you earn.

Clubs can choose to have their own unique call signs; however, they have some rules to do so. The club must be comprised of at least four members and they can be asked to verify their existence any time by the FCC. The FCC can ask them to produce documents showing they have more than four members, conduct meetings and so on. The club license is granted only to individuals who are known as "trustees" of the club station. Any club or licensed amateur can get a call sign for a short period of time for an event that holds significance to the amateur community. Special event signs have one-lettered prefix and suffix. The club trustee can apply for a call signal by contacting a club station call sign administrator, and the request for special event call sign is raised with a special event call sign coordinator.

Let us now look at the examinations in the volunteer licensing system.

Licensing Examinations

While the FCC takes care of military and commercial license applications, it is not responsible for amateur radio licensing examinations. Ham radio license exams are run/administered by volunteer examiners in the US—they are certified experts from a coordinating organization known as the volunteer examiner coordinator (VEC). These examiners administer the amateur license tests and, for each test, least three VEs must be present at the location where the test is conducted. To be able to give the test, they should hold a license with a class higher than the prospective license for which they are giving the test. For instance, to give a technician-class exam, the examiner must hold at least a general class license. If the VE holds just the technician license, he can still help in the process by assisting in running the sessions (excluding the grading and handling exams for higher classes).

The VEs take care of all the paperwork and even file the examination results with the FCC. They are responsible for various aspects of the test, including announcing the test sessions. VEs also ensure all the paperwork is in place. However, the final result is still granted by the FCC. The list of VEs, details of other resources and the test sessions are maintained by the coordinator. In addition to this, they also help in getting your license renewed and update your name/address details as necessary.

Any amateur holding the license can opt to become a VE by getting in touch with the VEC organization and completing the process needed for becoming a VE.

To take the exam, follow these easy steps:

1. Contact a local VEC and find a test session. You can even check online to find sessions registered with VEC/ARRL on the licensing website. While there are certain sessions that are open to public, there are some that are held occasionally, with access to certain group of people only.

2. Once a session is selected, see how to register—if it can be done online or whether they only accept walk-ins. Reach the venue on time and pay for the exam.

   Your identification will be verified by the VEs and only then you will be allowed to sit for the test. It is comprised of 35 multiple-choice questions. To pass the exam, you have to get at least 26 questions correctly.

3. Once you complete your test, it will be checked by the VEs. If you don't pass it, you can take it again. If you do pass, you will be asked to fill out two forms—the NCVEC Quick Form 605 and a Certificate of Successful Completion of Examination (CSCE), which is your temporary receipt of passing the test. Once the grades are updated in FCC database, you do not keep the CSCE anymore as all the information can be found in the database.

Next, you need to find your call sign. Before, hams had to wait for weeks, and sometimes even months, to find their call sign and to get the notice saying they had been upgraded to the next class. But in today's high-tech world, where everything is available online, hams do not have to wait for more than a few days. They can check the FCC database to find all the related information, including the call sign. All you need to do is log into the FCC's Universal Licensing System and click on *Licenses* (found next to the search button). Under Service Specific Search, click on *Amateur* to see the Amateur License Search page. You can enter your last name and zip code and click on the search button, by scrolling down the page. You can also see the Call Sign or Name Search service on the ARRL link. When your license is issued, you will be able to see your name with a brand-new call sign next to it.

Preparing for the Examination

You have decided to get your ham license and you even know the process—how the exams are conducted, how it is taken and administered by VEs. Now you must prepare to take the exam. Knowing the format of the test and how to prepare for it comes in handy. After all, getting licensed is not easy, so having pointers on how best to prepare is crucial.

The first thing is to demystify the test—how it is designed and what it is comprised of. As we now know, the tests for all these license class are multiple choice. You do not have any essay-writing questions, or questions based on scenarios. You are not required to write any stories.

For each license class, you have a test which is known as an element. And then there is a five-words-per-minute Morse code test, known as Element 1, which tests your knowledge on receiving Morse codes. Then there is a written test to obtain the technician license, termed Element 2, and the written exams for general and amateur extra class are known as Elements 3 and Elements 4, respectively. All the questions that you see in the test are from a question pool which is available to the public. It is a large set of actual questions that are used in the exam, which means the questions you see in your test are from this pool of questions only. They cover four domains—

- **Rules and regulations:** Rules you should know in order to operate legally.

- **Operating:** Basic operating procedures a ham should follow when on air.
- **Basic electronics**: Covers concepts around electronics and radio waves.
- **RF safety:** Covers concepts around how to install and operate transmitters and antennas, ensuring safety.

The test is comprised of random number of questions from these four areas, however, altogether there are 35 questions in the technician test, 35 questions in the general test, and 50 in the amateur extra test. If you get three-quarters of all the questions correctly, you pass the test.

If you want material to study, there are several licensing study guide books that are created especially for people studying for the technician exam and for other classes. In addition to these guides and eBooks, experts feel it is better to study in a group as that leads to better learning. There are classes conducted to teach the topics—these classes can be found by looking for information online, or asking at a ham radio. Other ways of finding these classes include schools and colleges, organizations that offer space for conducting classes and safety agencies.

Mastering Morse code depends on the likes and interests of an individual but it can result in great sense of pride and accomplishment. The style that most hams know is the Farnsworth Method. While sending the characters, normally the dits and dahs of each character are sent at the code speed you want. However, the characters are spaced in time, which means the overall word speed is not enough to process the sound pattern of the character. Most beginners send the characters at seven-words-per-minute (wpm), but the overall speed is reduced due to character spacing for the words. If the characters are sent at a higher speed, you need to look up the words and you can do that only for a few seconds per minute. Instead of the "dot dash dash" pattern, the Farnsworth method gets you to send "dihadhdah," which is closer to the sound of the character. As part of this, you don't need to memorize the characters as disconnected elements, rather the entire pattern of sounds can be memorized. With this method, you do not have to remember the tables and hence you can process faster and easier.

An important technique in mastering Morse code is to just keep practicing as it will help you learn and conquer new letters easily, at higher speeds. So you just need to overcome the initial barrier when your brain is busy with new wiring. And as you come across new codes, practice them in your daily life so that your brain gets acquainted with them.

License Renewal Process

With all the effort and hard work, you earned yourself the amateur license, good for a 10-year term. The good news is that you do not have to take any test again to get your license renewed at the end of this 10-year period; you can simply visit the FCC website and renew your license. If not online, you need to fill out the FCC Form 605 at least 90 days before the expiration date and mail it to the FCC (the 90-day period is needed to prevent license expiration). And if the license expires, you will have to stop transmitting but will have a grace period of two years to re-apply for it without taking the exam. If you lose your license or it gets damaged, you can request that the FCC issue you a new one. All you need is a letter stating you have asked for a replacement.

Responsibilities as Part of Licensed Ham Radio

Being a licensed amateur radio, you must remember your primary responsibilities at all times and ensure that you operate your station in accordance with the rules set by the FCC.

Unauthorized operation: This rule set by the FCC prevents the use of improper station equipment when you are not available. For example, your spouse isn't allowed to operate the station in your absence since he/she doesn't have the license to use it and you are not around to ensure proper operation. Even though they know how exactly it works and they operate with you frequently, they shouldn't do it in your absence. You can adopt one of the many ways available to ensure the station is not operated by any unauthorized individual, such as disconnecting the microphone when you are not present.

Personal information: The FCC wants all licensed hams to enter their valid email addresses and other details in the database so that it is easy to contact any ham in case of emergency. In case any ham changes their address or other details, it should also be updated in the database. Other important information that is held in the database is the Federal Registration Number (FRN), which is the identification number assigned to you as a licensed user.

Station inspection: As part of your license, you are required to make your station available so that it can be inspected by an FCC representative, if need be. This comes as part of the terms and conditions you agree to. Although this doesn't happen too frequently, always remember to keep your license available at all times.

Privileges and Bands

The frequencies, methods and modes that hams use, as per the rules and regulations provided by the FCC, are known as *privileges*. The Communications Act of 1934 gave the FFC authority to grant privileges to hams through the licenses. So, when you sign Form 605 and apply for the license of a specific class, it means you agree to adhere to the FCC rules, which agree to stay within the granted privileges of your license.

The most important privileges granted to hams are frequency privileges. Back in the day, when the world was first introduced to the concept of radio, hams, as well as military and commercial stations, used to use one huge band, which resulted in crowding of this big band, leading to several issues. This made the authorities decide that hams should be allocated certain worthless frequencies for their specific purposes. This worked great for hams as these frequencies enabled them to communicate over long distances. However, this happy situation didn't last long as the shortwave bands were used up by different users and hams saved the first bands for themselves. Today there are a huge number of bands and various different types of radio spectrum users.

The frequency privileges assigned to various services are known as *allocations*. For instance, amateurs are granted access to 144—148 MHz, which is the two-meter band. The amateur allocations are widely spread across the radio spectrum; they are not concentrated or excluded from a specific region. As we know about radio signals, these signals, when transmitted over different frequencies, act differently. Thus, it is good news that amateurs have access to a wide range of frequencies so that they can be used differently for different communication purposes. Technician-class licensees are given access to six-meter bands (50 to 54MHz), two-meter bands (144 to 148 MHz) and 70-cm bands (420 to 450 MHz). With the amateur HF bands, the frequency granted to a ham is something that is decided by the license class. When a technician-class licensee is upgraded to higher classes, he is granted access to more privileges.

Within the bands granted to hams, there are certain additional restrictions that are made based on emission type. Emission refers to any radio signal given out by transmitter. While frequency privilege is permission granted to use certain frequency, an emission privilege is permission granted to communicate using a specific mode, for example image, data, etc.

Now, when license class, frequency and emission privileges are combined, it they lead to a division of ham bands into sub-bands. The specific parts of these bands, which allow for the use of only certain modes, are termed *mode-restricted*. The reason for having these mode-restricted sub-bands within the amateur bands is that the methods of operating different modes is, at times, not compatible. For instance, phone and data operation happen differently and there is interference between the signals. And with the increasing number of modes, there are more sub-bands created with the amateur bands. This way, hams use narrow-bandwidth modes, like CW, at low frequency side of the bands, and wider-bandwidth modes, such as data, in the higher frequency side of the band.

Coming back to allocations, it would be amazing if each radio service had its own exclusive access to allocations. For example, we have exclusive bands for amateur radio services across the world. But we cannot have separate allocations for each service, considering the fact that the spectrum is limited, and services have their own specific communication needs. Therefore, two services have to share the band, which is known as shared allocations, such as some amateur radio bands. In such cases, one of the two groups is given priority and is known as primary allocation. The other group is given less priority and is known as secondary allocation. That primary allocations group is known as the primary service, and the other one is called the secondary service.

Since two allocations coexist, the primary service is protected against the interference by signals from the secondary service, which gets access to those frequencies with an understanding that it cannot cause any harm to the primary service. For instance, amateurs belong to the secondary service in the band of 70 cm and hence should not interfere or cause harm to radiolocation stations that fall in the primary services category. This way the frequencies are available to more users as compared to a scenario when frequencies are uniquely assigned to each service. And due to this reason, hams today share several bands and have access to a wider range of frequencies.

This way, all the UHF and other higher bands share frequencies in some sort of pattern, which might be applicable to some geographic areas or military space.

Canada has assigned 420-430 MHz to other services and, to prevent international interfaces, US amateurs are told by the FCC not to use the segment falling near the Canadian border. While the amateur HF band is mostly not shared, the 40 meter is shared with shortwave stations and the 60 meter is shared with the US government. This doesn't mean all the amateur allocations are allocated to amateurs; when there are other services that are on higher priority, amateur service becomes secondary. To know about the allocations, it is good to read Part 97.303, available on the ARRL website, to avoid interfering with other frequencies.

Let's understand the concept of a Band Plan now. Finding out about different activities becomes simpler if you have an understanding of band plans that organize activities by these frequencies. This enables you to know about the specific type of activities, which helps in grouping the similar ones together so that the spectrum can be used effectively. However, it is interesting to know that the band plans are not created by the FCC; these are created by the amateurs themselves (plans can be found on the ARRL website). So, in simple words, band plans include temporarily grouping together different activities by amateurs, to ensure optimal utilization of the spectrum. Although they are not FCC rules, they are considered to be good practice to adhere to. The plans are applicable during normal conditions and do not guarantee using any specific frequency at a given time. This means amateurs can be flexible with the frequencies, except for repeater frequencies. But an important thing here is to know how it is decided which repeater will use specific frequencies. Even this is decided by the amateurs themselves; the FCC has no role to play here. Amateurs have come up with a system of regional frequency coordination to ensure the bands are used effectively by repeaters and interference is avoided as much as possible.

One or two segments of the available bands are comprised of groups created by grouping repeaters and their auxiliary stations. Each group has fixed input and output frequencies, and a common offset. This ensures limited spectrum with maximum repeaters. A committee of volunteers, called frequency coordinators, suggest the frequencies for reception and transmission, and wherever they find regions to be overlapping, they work towards reducing the interference between the frequencies as much as possible, ensuring orderly coordination.

International Rules and Operating

Once, every country used to make its own rules for ham radio, but with the advancement in technology and equipment industry, the world felt the need for coordination and consistency. Therefore, the governments joined hands after World War I and started creating treaties on how countries should regulate communications. It all went well till World War II, when the communication industry made further advances. As a result of this, there emerged the need to discover new ways of radio communication.

International Telecommunication Union (ITU): The FFC has authority only in the US, and similarly, each country has its own equivalent agency. So the question here is of coordination between these different agencies. If every agency starts making its own allocation, it will be total chaos, as radio waves are not different for different countries. To avoid the chaos, international coordination is a must, and therefore the ITU was formed as an agency of the United Nations. The main aim of this administrative forum is to develop international telecommunication laws and treaties and frequency allocations. It also maintains the international radio laws that all the United Nations countries should adhere to.

To ensure proper coordination between different parts of the world, the ITU has divided the entire world into three regions—regions 1, 2 and 3. This helps in organizing the allocations more effectively. Since VHF and UHF travel mostly till the radio boundaries, it was decided that the allocations that fall in these bands do not have to be precisely aligned. Due to this, the amateur allocations are spread across multiple regions and are not limited to just one. There are allocations for amateur radio service in regions 1, 2 and 3. On the other hand, HF allocations maintain more consistency between the regions, however, there are still some variations for US amateurs.

Irrespective of the ownership of an aircraft or citizenship, the radio rules change at the boundaries of these defined regions. For example, if you are handling maritime mobile service, when you move from region 1 to region 2 you will have to follow the rules meant for region 2. This holds good even if you are a citizen of region 1.

Let's discuss the international operating part now. Operating in a different country can be a unique experience altogether. You get to meet new people and can attract crowds from that country (if you are licensed to use the HF band) by becoming the DX. However, to operate in a foreign country you need permission for amateur operation. Also, the country should permit the amateur operation. If you are holding a US amateur license, to operate in a foreign country you must have reciprocal agreement between the two countries. You can get this operating permission three ways—an International Amateur Radio Permit (IARP), reciprocal operating authority or by the European Conference of Postal and Telecommunications Administrations (CEPT) agreement.

International Amateur Radio Permit (IARP): IARP is a body that allows US amateurs to operate in certain South and Central American countries without need of any special license to enter or operate from that country.

Reciprocal operating authority: Several countries have entered into reciprocal operating authority agreements with the US to ensure they recognize each other's amateur licenses. In simple words, it is a government vs. government agreement that recognizes each other's license.

European Conference of Postal and Telecommunications Administrations (CEPT): This agreement allows the general and extra class license holders to travel and operate from most European countries without any need of an extra permit or license. All that is needed to travel to and operate from a CEPT country is your original license, a copy of the FCC's notice about CEPT and proof of your US citizenship.

Chapter 10—Operating Regulations

It might be good news for some and bad for others, but the reality is there aren't too many regulations around operating amateur radio. At least for certain things, the FCC likes to rely on the licensed hams, as long as they meet the technical requirements and basic rules are followed. The hams create their own procedures and regulations, and the FCC comes into the picture only when there are intractable issues or disputes to be resolved. This gives hams liberty to adapt the rules and operate independently. However, all amateur transmissions must be done with consent of a control operator, who is a licensed operator responsible for ensuring all the rules set by the FCC are followed in the process. For a station, there can be only one control operator at a time.

A control operator plays an important role in the entire process:

- He is responsible for creating a signal.
- He is an amateur responsible for ensuring all the transmissions adhere to the FCC's rules. He doesn't have to be the station licensee, but the station licensee designates the control operator. The control operator doesn't have to be present in person at the transmitter, but all the transmissions are his responsibility.
- The details of a control operator must be fed into the FCC database. This is necessary because the FCC has the right to know who you are, whether you are licensed and what your contact details are.
- The control point is generally at the transmitter, and the control operator manipulates the transmitter controls.

Guest Operating

Being the control operator, you have certain privileges; being the station owner, you have others. However, if you are operating the station, you adhere to the privileges granted to a control operator—it's simple. However, in case of a guest operation, you might either be allowing the other person to use your station or you might be the guest yourself. Whatever the case, it is important to know what sets the privileges for control operator. A guest operator hosted by a higher-class licensee can operate using the privileges of the host only in cases where the host is the control operator. If that's not the case, the guest has privileges only as per their license. Let us try to understand this with the help of an example—you are, say, a technician-class licensee and are invited to Elmer's station, who is an extra-class licensee. When Elmer is the control operator, you are allowed to operate the station on any of the amateur bands available. But when Elmer is not around, you only have your technician-class privileges.

In a case where you are the guest and have a license that is higher than that of the host, you can use your higher-class privileges to be the control operator, irrespective of whether the host is around or not.

No matter what their license class is, both the station owner and the guest operator are responsible for operation of the station. And the control operator is responsible for the transmissions. The station owner limits the access to the station only to licensees who can follow the FCC's rules. But there can be times when an unlicensed operator ends up operating the station. Consider a scenario when you visit a club and make a few contacts, or when you just pull the microphone off your ham friend to say hello to another friend over the air; you're acting as an unlicensed operator, but under complete supervision of a control operator. There is no issue with these scenarios where the unlicensed operator uses an amateur station since the control operator is around when transmissions are happening. But if the licensed control operator is not around and is somewhere else, busy with his or her personal chores, family members cannot operate the station since they don't have the license to do so. It is important to keep in mind that the control operator must be present at the station when transmissions are done so that he can ensure the FCC rules are followed, no matter what! But if unlicensed operators want to use the equipment just to receive messages, that can be done.

Identification Rules and Process

In most cases, the other person cannot see you when communicating. And in all these times, there is no way, other than using the call sign, to identify a signal. After all, the other person should know who is transmitting from the other side, and when you are on the air, your call sign is your identity. These identification rules are applicable when you are transmitting from home, a forest or from the top of a hill. The identification process is important not just for the FCC to know who is transmitting, but also for the other ham who is trying to contact you. Due to this, identification is one of the things the FCC handles directly.

When it comes to the identification process, the first and foremost rule is that unidentified transmissions are not allowed. By unidentified, we mean a transmission that has no call sign associated with it. Even if you want to make a short transmission to test your equipment or make adjustments, just stating your call sign should do the job. For example, in case you want to know if you are around a distant repeater, don't plug in your microphone to listen to the signal from the repeater. This is known as *kerchunking,* as you will be listening to the sound that is being monitored by everyone. You should just state your call sign.

So, once it is confirmed that the transmission is not unidentified, we can proceed with the identification process, keeping the simple rules in mind.

- State your call sign at least once every 10 minutes after the connect is established, till the communication is over. You have to mention it to establish the connect. And if the connect lasts for less than 10 minutes, state the call sign before you end.
- You do not have to state the call sign of the other station as the whole idea of the identification process is to identify you and not others. So only state your unique call sign. But if you are stating the call sign of other station, it is considered a good practice. In case you are stating both, just mention "you ID" whenever you are stating your call sign and "For ID" for other stations.
- You can also state your call sign in Morse code, in an image, by voice or as part of the digital transmission. If it is in the form of a digital or video call sign, ensure that it is in a format that everyone can receive. If you are using your phone to send out your sign, ensure you are identified in English, irrespective of the language you are using to communicate.

Then we have something known as tactical call signs, which are used to identify the location and purpose of a station. Some examples of these types of call signs include "first-aid station," "fire watch on Golden Bridge" and so on. Although these can be used anytime, they are mostly used in case of public service operation when communicating or during emergency. However, one shouldn't get confused with tactical call signs, thinking these are replacing regular call signs. But the regular identification rules are applicable for the tactical call signs as well—you have to state your call sign once in every 10 minutes, to start the contact and before you end the communication. The main aim of a tactical call sign is that it allows for a consistent identification process to ensure communication is rationalized based on the function. Imagine how it would be if everyone had to remember which call sign was used for which function.

Another attribute that helps with the normal identification process is the self-assigned indicator. Whenever you are operating from a location away from home, you should add certain types of information to your sign so that other hams can be aware of your location. For instance, a station will add some information to its call sign when operating from other states. If this is not done, the special prefixes used for this state can lead to confusion about its location. Say, if NL7CC is operating from a location that falls in another region, the ham can add extra information, like /W3, to state he is operating from the third region. In this case, his call sign becomes NL7CC/W3, where W3 is the self-assigned indicator.

According to part 97.119(c), a call sign should have one or more indicators included, and these must be separated from the call signs so that the distinction can be clearly understood. However, care must be taken while selecting the indicators as they shouldn't conflict with any other indicator mentioned as part of the FCC rule; they should not conflict with the prefixes and suffixes allocated to any countries.

Another indicator that is used in amateur radio services is the upgrade indicator. Once you receive the license for a specific class, and if you are really interested in doing well in amateur radio services, you will certainly want to upgrade to the next class, which means you will want to pass the examination for the next level. And once you clear the license for the next class, you will be granted new, advanced privileges and this will be reflected in your call sign by adding appropriate indicators to it. By adding the additional two-letter indicator, the FCC lets others who are receiving your transmission know that you have upgraded your class license. Even if the information is not updated in the database, your call sign will speak to itself.

Below are the indicators that are added to the call sign to inform others of your upgrade:

- Upgraded from novice to technician class: add "/KT"

- Upgraded from technician to general: add "/AG"
- Upgraded from general/advanced to extra: add "/AE"

Identification rules are applicable in case of on-the-air test transmissions. Following the same basic rules of using call signs, these must be stated once in every 10 minutes and towards the end of a contact. In case of a test transmission, one must ensure that they are kept brief so that there is no interference with the other frequencies, and they should be kept from occupying a useful frequency. Normally, for test transmissions, this is how the identification happens: "W1AW testing".

Stations that are under automatic control should also adhere to basic identification rules and use devices and procedures that are in line with the FCC rules. One of the common types of station that uses automatic control is repeaters. These should identify themselves using the call sign of the station, by voice, image, standard video or by Morse code.

Another case is special event stations, when an amateur club arranges to get the license to use special event call sign. In such cases, both the regular sign and the special call sign must be given on the air for transmission. This helps the listeners in identifying the station that is sending special event transmissions.

Let us know look at the rules and call sign changes in case of guest operating.

When you are moving from your station to another station, you need to inform users who are receiving your transmission about the change in your location, unless your host is lending you the station. You must identify yourself using the call sign of your host. If you, being the guest operator, have a higher-class license as compared to the host of the station, you can identify yourself with the host's call sign, followed by your own sign. For example, if you are holding a general class license, having a call sign KD7FYX moves to another station whose host holds a technician-class license with call sign KD7PFA on the voice band, where technicians do not have privileges, you call using the call sign as KD7PFA/KD7FYX. However, if the transmission has to go on the VHF band, you can use the station host's call sign, that is KD7PFA, as both his licenses have privileges to access the band.

It is quite logical, if you think of it from receiving point of view. If the listener hears it as KD7PFA on a band that he is not allowed to transmit on, and looks up the call sign in the database, the receiving station will think it was transmitted using a frequency that doesn't have privileges to do so, being a technician-class licensee. But when the call sign includes the call sign of a higher class that has the required privileges, it provides a clearer picture.

We are now aware of the identification rules; however, there are two exceptions to these rules:

- Remote-control signals and
- signals retransmitted through space stations.

Since remote-control signals are weak and cannot traverse long distances, you do not need to use your call sign. And if you are transmitting signals through space stations, you don't need a call sign either. This is because space stations are located more than 50 kms above the surface of the earth so there is no need to identify yourself, as the station will use its own call sign for transmissions. Even for the ham satellite transponders that receive frequencies and retransmit on a different band, you are not required to use your call sign.

Interference in Transmissions

We have been talking about interference and overlapping of frequencies. Let us dive deeper and see what causes interference. Interference, in simple words, is the noise or disturbance caused by various attributes—natural resources, unwanted signals radiated by devices, equipment, appliances, etc. For now, let us look at the interference caused due to unwanted signals from other transmitting devices.

Interference caused in the signal due to nearby signals is the price we pay for having flexible frequency. If the transmission happens on the allocated band, or defined frequencies, it's regarded as minimal, and even these defined channels are overloaded, not leaving enough bandwidth for the unwanted signals to barge in. Although interference is not illegal, it causes too much inconvenience and disturbance. Though most interferences can be managed, hams have also learned different ways to handle them:

- equipping the radio with filters that are capable of rejecting the interference
- remembering that no one owns these frequencies, not behaving as if you do and planning accordingly
- courtesy can help solve most of the issues
- taking into consideration all sorts of activities

If a transmission seems to obstruct or interrupt a communication service, it is known as harmful interference. Every amateur should work towards ensuring there is minimal harmful interface while transmitting or receiving. You should also keep an eye on reports that highlight spurious signals or interfering frequencies. You should always do a short test transmission by using a dummy load.

Even after paying attention to all the details, there can be accidental interference, for instance—atmospheric conditions or the ionosphere can lead to changes in propagation on a band. It can cause a signal to strengthen out of nowhere, causing disruption in your connection. Sometimes an amateur might start listening during the pause time and start transmitting, thinking what he heard was for him. Even changing the direction of antenna can lead to strengthening a spurious signal, which can lead to interference. So such accidents do happen and therefore, you cannot always expect to have a clear frequency band. No one owns the frequency; be flexible and tune your frequency a bit, if it happens. You can also consider moving your antenna or reducing power to deal with it.

There are only certain circumstances under which the interference is prohibited as per the FCC rules; for example, when interference impacts the functions of critical services such as radio navigation. For other conditions, don't overreact and understand that it is not always intentional. The receivers we use today are sensitive; they cannot be expected to do their best while rejecting the strong signals that come by their way. Moreover, even the signal processing features, such as preamplifiers and noise blankers, can lead to issues.

Noise blankers are a frequent source of problems as most of them operate on sensing short noise pulses. The way they work is by looking at the entire frequency spectrum and not just at what comes their way through the signal filter. The moment a noise blanker senses a strong signal around it, it can totally get confused, to a level that it can lead to receiver shutdown or cause splatters over many kilohertz. Therefore, it is better to turn off the noise blanker, especially if the exposed band is full of strong signals.

You can deal with this situation by using RF attenuators instead of noise blankers, if you have strong signals nearby. As the name suggests, attenuators tend to reduce the strength of strong signals that try to overload the receivers. Overloaded receivers are very capable of producing spurious signals and bringing the noise up and down the band. Therefore, attenuators help in increasing the signal-to-noise ratio—it cleans up the band.

And then there is the RF gain control that can make the receiver sensitive, but at the same time, reduces its tendency to get overloaded. You can always experiment with the RF gain control on a busy band to see if it improves the efficiency of your receiver.

If you are using a variable bandwidth, passband tuning or similar controls, they will also help you in seeing cleaner band. So, if you have any of these digital signal processing (DSP) features, have another friend help you achieve a quieter band.

This discussion was all about accidental interference and the way you can handle it. But there are also times when the harmful interference is intentionally created, and this is known as willful interference. Willful interference is prohibited by the FCC.

Different Types of Transmissions and Communications

Third-Party Communications—The Hows and Whys

Ham radio is used for communication purposes by licensed people and organizations. One of the activities conducted by hams, which is quite popular, is third-party communication. As part of this, a message is sent by a ham radio, which is then relayed from station to station, till it reaches the destined station. Since, in this process, the regular telephone and other systems of communications are bypassed, various foreign bodies are interested in controlling it. However, the FCC doesn't want it to become a non-commercial system, and therefore there are some rules for third-party communications.

Before we get into details of the rules, let's first define the important aspects of this type of communication:

- The entity who wants to send the message is known as the third party, and the control operators on the two sides, who make the radio contact, are the first and second party. A third party can also be the entity that wants to receive certain message using this system.
- A licensed amateur, who can become the control operator at either of the stations, cannot be considered as the third party.
- The third party doesn't have to be present in either the transmission or receiving station. The message can be taken from the person to the ham station or can be transmitted using speech over the phone, to the ham radio (a process known as phone patch).
- The messages transmitted by the third party don't have to be in written format. Images, spoken words or data—all can be third-party communications.
- The third party can participate in receiving or transmitting the message at either of the two stations. When an unlicensed person at a station wants to send third-party communications, they can speak into the microphone, type on a keyboard or send Morse code
- Even organizations, such as schools or churches, can be third parties.

So, if we want to define third-party communication in simple words, it is the communication as part of which an unlicensed person or an organization sends or receives information via ham radio, even though the person who wants to send or receive the message is present at the station.

FCC considers third-party communications as vital activities conducted over ham radio, and therefore hams are trained to provide efficient third-party communication. Live conversation and phone patches are all part of both emergency and normal communications. Hence, third-party communication can happen between the amateur stations that adhere to FCC rules, with the constraint that the communication should not be commercial. However, when the signals cross the defined borders, a change occurs in the rules defined. Since not all countries allow third-party communication, it is restricted to those that allow third-party communication with US-licensed amateurs. If a country is not in agreement with the US for third-party communication, none of the third parties can have a conversation with them.

To make the definition of third-party communications clearer, let us look at the following points:

(a) Letting your relative or neighbor (an unlicensed person) make a contact when you are around is known as third-party communication, irrespective of the duration and purpose of the communication.
(b) A message transmitted from one ham to another is not considered to be a third-party communication, irrespective of whether the message is transmitted directly or relayed from one station to another.
(c) When a contact is made with the help of hams to connect a student staying in South America to his parents, this is considered a third-party communication, even when both the parties are present at the receiving and sending stations. The message can be taken from them in any form—speech, images, etc.
(d) If you connect with the DX station to convey a message to your family in your state, it is known as a third-party communication. Ensure the DX's station country has an agreement with the US to have third-party communication.

Prohibited Transmissions

There are only few types of transmissions that are prohibited as hams have access to a huge band to communicate with others, adhering to procedural and technical rules. While other services have strict regulations associated with them, hams are always encouraged to learn new things by experimenting. The following transmissions are the only ones prohibited, for obvious reasons:

- Deceptive or false signals: These transmissions are meant to deceive the receiver, for example—using someone else's call sign.
- Unidentified transmission—These transmissions do not have any associated call sign during the required time.
- Obscene or indecent speech—These are signals that contain topics to be avoided
- False distress or emergency signals—These are signals that do not require the attention they are given by the FCC and other authorities.

Apart from this, even the regular communications that can be performed using other stations are prohibited from being transmitted over ham radio. For example, the transmissions needed to direct the traffic on a lake should be done on maritime or local stations. Since none of these transmissions help in keeping the amateur radio useful, they should be prohibited at all times. Let us now look at some types of transmissions that do not happen over amateur radio:

1. **Business communication:** According to authorities, no business communication can happen over amateur radio. After all, it is amateur radio, which is meant for certain special purposes, and hence shouldn't be used for commercial services. However, the authorities do not consider someone's personal activities as part of business communication and therefore, ham radio can be used for those. For example, you can use a ham radio band to talk to your family members in case you are not able to reach them directly. You can even use it to order things online, as long as you stick to the FCC rules meant for amateur radio usage. It is totally okay to advertise your products on this channel as long as it is not part of your regular business. Below are some of the activities that can and cannot be performed using amateur radio band:
    a. Permissible activities – advertising your product, using a repeater's autopatch to change a doctor's appointment, describing your business in conversation.
    b. Non-permissible activities – selling sports goods online, advertising your professional services, using a repeater's autopatch to call a client.
2. **Encrypted transmission:** Amateur radio service is known to be a form of public communication, and we agree to this as part of the agreement we sign for using ham radio services. As a result of this, we cannot transmit encoded signals or encrypted codes if the intention is to prevent others from receiving the transmitted information. When the information is translated into data, it is known as encoding, and recovering the information from this encoded data is known as decoding. Although most forms of encoding are okay to be used for transmission if done using an allowed protocol, hams have the authority to look up the protocol and can create the capabilities to decode the sent data any time they want. However, the process of encoding that uses codes to hide the information is known as encryption. And when the information is recovered from the

encrypted data, it is called decryption. For amateur radio services, encryption can be used only for radio control or for transmitting messages to space, where interception can lead to serious issues (hence, the information is kept secured by encryption).

Not everyone understands the difference between encoding and encryption. And, knowingly or unknowingly, if users encode the information using a protocol that cannot be understood by hams, it is considered encryption and not encoding. But if the protocol is available to the public, the transmission is known to be encoded and is therefore permitted to be transmitted using amateur radio. The basic rule here is no information should be hidden from hams.

3. **Broadcasting information and retransmission:** The general public often refers to ham transmission as broadcasting; however, it is not correct to call it so. Broadcasting refers to the one-way transmission that is meant for the general public, and hams are not allowed to make such transmissions except for the times when they want to transmit information bulletins, code practice or provide information needed for emergency situations. What's prohibited on broadcasting is transmissions such as relaying information from other services, etc.

On the other hand, hams are not allowed to assist or transmit any information related to news gathering by broadcasting firms. Transmitting music is also prohibited on amateur radio, including incidental retransmission of music from a radio located nearby. This means that if you are using ham radio in a space where there is music, you should turn down the volume of music so that it is not retransmitted over ham radio. According to the rules defined by authorities, music can only be retransmitted as an authorized rebroadcast of transmissions from space stations.

Even retransmitting the signals generated from another station is not allowed on ham radio, except for the cases when the digital data or messages are being relayed from another station.

Although ham communication is meant to be transmitted and received only by hams, there are certain exceptions from the rules set for broadcasting.

- Hams can retransmit propagation and weather information from the government stations, but only when there is a strong need.

- Hams are often seen operating from unusual places. They also operate from boats or planes, for which the details are covered in Part 97.11. Hams can operate from aircrafts but only with the approval of the captain as there are FAA bans on transmitting from inside the aircraft for all commercial passenger flights. And if you get the approval from the captain, you need to have your own radio equipment; you cannot use the radio of the aircraft or boat. Also, amateur signals cannot interfere with transmissions from any other radio system that is being used on-board as it can cause issues in the navigation system of the aircraft or boat.
- As a general rule, hams cannot have any sort of communicate with non-amateur services; however, the authorities might let hams talk to non-ham services during declared cases of emergency or certain other times.

QSL Cards

Another fun element about ham radio is collecting QSL cards. Amateurs collect QSL cards from other hams that they talk to on the radio. Just the way some people are crazy about collecting stamps and coins, some hams like to collect QSL cards. As stamp collectors always look for interesting stamps on the envelopes they receive, QSL card collectors anxiously wait for these cards. While a hobby is certainly a reason to collect these cards, getting a chance to participate in the certificate programs is another reason. If you wish to give these cards to your contacts or/and friends, you require a personal QSL card. You can be creative and design your own. Or, you can get it printed or order one from one of the many vendors available. Think about what you want to get printed and get it done accordingly. Also consider the quantity that you would need to order. The content can be something like your call sign, a placeholder to include important details (RST, call of the station you had contacted, date and time of the station, mode, etc.), your name and full address, the country you are in, grid location and so on.

To send a QSL to a ham located in another station, you have two options—you can send a QSL directly through a post office or send cards through a QSL bureau. The second option, using the bureau service, is considered to be the most effective and least expensive method of collecting the cards.

Chapter 11—Use Communication Skills to Share and Care

It is all right to set up your ham radio station with the intent of having fun or indulging a long-cherished hobby. However, as you become more and more experienced with your work, you will feel the need to use your communication skills in ways that are more practical. In fact, during the journey from amateur to expert, you will wonder if you should not be doing some kind of community service too. After all, you have an approved license to handle all manner of radio frequencies, enforce diverse operating rules and receive protection from varied types of interference. Best of all, you have become capable of maintaining high technical standards. Therefore, you may strive to use your training and expertise to handle communication pathways during emergencies too. At the same time, do not forget to go in for training exercises, keep a keen eye on changing weather conditions and provide support at sporting events and public functions. To sum things up, by using your wonderfully extensive network 24/7/365, you handle both ordinary calls and emergency ones too. This way, you give back to amateur services.

A. Linking Up with an Emergency Services Establishment

An emergency can be anything, such as earthquakes, wildfires, extreme weather conditions, a collection of debris from broken space shuttles and so on. Whatever it is, you must prepare yourself to tackle everything, right from the moment you obtain your license. Ham radio operators refer to emergency communication services as EmComm. The idea is to ensure that communication services suffice to immediately lessen threats to life or property. Towards this end, people may expect you to report everything from relief efforts undertaken during the occurrences of natural disasters to major vehicular accidents and so on. You are not to judge what is important and what is not. Your job is to disseminate the requisite news and information.

ARRL

You may have an interest in EmComm merely to support your family and yourself. Alternatively, you might want to know how ham amateurs prepare themselves to participate in organized EmComm. Whatever your reasons may be, you ought to know relevant information about where to tune in and how to interact with other amateurs. In this kind of a scenario, the ARRL has a great role to play. Unlike the regional and local EmComm establishments, the Amateur Radio Emergency Service (ARES) of this field organization is highly popular and has a large fan following. Of course, this does not mean that the regional and local communities are not active. They are, but ARES has the advantage of individual 'bosses' for individual sections. These controllers of diverse sections may be a few counties functioning as a group, or an entire state. The differentiation is dependent upon the size of the populace. This benefit has an association with an ARRL membership. The Federal Management Agency (FEMA) is responsible for organizing and managing ARES and RACES (Radio Amateur Civil Emergency Service). These establishments link to communications, which deal with national emergencies. They provide assistance to both private and public agencies. It does not matter if the emergency is a natural disaster or a civil one. However, they are open to amateur operators like you. Thus, you are welcome to request an appointment with ARRL's field organization whenever you feel a desire to administer or handle EmComm activities linked to your ARRL section. Just make sure that you garner some experience prior to your participation.

When you volunteer your services, you may fill any of the following positions.

- **Technical specialist (TS)** – You may opt to share your expertise if you have skills related to a certain aspect of ham radio operations. Alternatively, you may have extensive knowledge about a particular area. In case you do get the position, you will be a consultant for both local and regional ham radio stations. Even the ARRL will consult you at times.
- **Public information officer (PIO)** – It will be your job to establish a rapport with the regional and local media, such that you may let the public know about the EmComm and public services that ham radio operators offer. Additionally, you must strive to build relationships with diverse organizations and community leaders.

- **Assistant section manager (ASM)** – Even though the organization appoints its own section managers, there is nothing to stop you from becoming their assistant. Your tasks will vary in alignment with the particular section you associate with, but every assistant in every section must gather data from volunteer reports and analyze them. Another duty is to work with, as well as check into, regional and local nets. If you are lucky, a special task may come up, and the authorities may request you handle it on behalf of your section manager.
- **Official observer (OO)** – You will have to keep a vigilant eye on the activities of other services, who engage in spurious transmissions. Sometimes unlicensed operators make an entry into the ham radio arena. Report them to the concerned authorities. Furthermore, you must help your fellow hammers too. Sometimes they receive FCC notices because they have violated some rules. These rules may link to technical or operating irregularities.
- **Official emergency stations (OES)** – Local emergency coordinators will guide you regarding the particular actions that you will have to perform at certain times. In this case, you will only associate with those radio stations which are committed to performing EmComm work. As a result, you will gain an opportunity to take charge of logistics, detailed projects linked to operations or administrative matters.

These are just a few examples of what you may volunteer for with ARRL. There are many other jobs that you may find even more interesting, and more connected with your skills and knowledge. The choice is all yours!

Benefits of being a part of ARRL:

QST: *QST* is the source of information and news around amateur radio. It is a journal that can be delivered to your desktop (soft copy) or doorstep (hard copy). So, if you become a member of ARRL, you get monthly subscription of this journal that gives you details about:

- Monthly hamfest calendar and conventions
- Informative product reviews of the accessories and radio
- Various kinds of articles covering information on broad spectrum of topics; some are easy to understand while others are challenging

- Eclectic technology, which is a monthly column that talks about emerging amateur technology and other commercial updates
- Public service efforts that amateurs are providing in and around the country. It also provides information on how you can join an aspect that interests you.

Web services (only for members): Being a part of the ARRL, you get services and information that are not easily available elsewhere. Some of these services include:

- Free e-newsletters that talk about different public services, radio clubs, ham radio news, various contests and so on
- The *QST* Digital Edition gets delivered each month in addition to receiving the *QST* by email. Members of the ARRL get an email every month with the link of the digital edition of the *QST*, which they can either download and save or read online. The new format used by the *QST* brings along with it several benefits such as timely access, enhanced content, interactive experience and much more. The digital edition is really interactive and easy to access; it is like a flipbook the pages of which can be flipped from front to back. It is also easy to navigate, search friendly, sharable and printable, easy to zoom in and zoom out, and is comprised of live hyperlinks. Each issue comes with certain unique features, both audio and video and additional content. However, the members are required to have login access to the current digital edition.
- *QST* archive and periodical search
- Product review archive that has copies from 1980 to present
- ARRL member directory that helps the amateur connect with other amateurs with the help of a searchable member directory. It shares photos, profiles and various other things with members who have similar interests.

Technical information service (TIS): You can reach out to the ARRL experts for all your technical or operational queries as this service is free for all ARRL members.

Affinity benefits and member discounts: Being a part of ARRL, you can avail yourself of special member discounts and ARRL-sponsored affinity benefits. The products and services offered by ARRL to their members are quite cost-effective and convenient. Some of these benefits include the Ham Radio Equipment Insurance Plan, the ARRL Visa Signature Card, Amazon Smile and Liberty Mutual Home and Auto Insurance.

Ham Radio Equipment Insurance Plan: Disasters can occur anytime, anywhere. No place is safe, and disaster doesn't knock at the door before coming in. And what is most helpful during these tough times is your insurance plan. With the ARRL Ham Radio Equipment Insurance Plan, you might not be able to get back whatever you had, but it can help you with what you need to go on the air. Insurance is needed to protect you from different types of damages and losses. If you damaged your mobile equipment, amateur station, antennas or anything else due to accident, lightning, flood, theft, etc., the Ham Radio Equipment Insurance Plan can help you out.

The ARRL Visa Signature Card: Now all amateurs can flaunt their ARRL Visa Signature Card as every ARRL member gets access to ARRL services and programs. The Visa signature card has no preset limit and members get 2500 bonus points with their first purchase, while redeeming the rewards and earning unlimited points. Some rewards include gift cards, merchandise and so on.

Outgoing QSL service: The ARRL handles all the overseas QSL chores and the savings that one can accumulate using this service can be used for various other things.

Education: All ARRL members can take courses that can help them to pass their license exam, upgrade themselves or learn more about amateur activities for public services or emergency communication. ARRL membership also offers videos and other content on operations, technical and licensing information.

ARES and RACES

Now, you may wonder why there are two organizations such as RACES and ARES in the same field. After all, both of them seem to be doing the same thing, and maybe they could make do with just one label/name! You are right, but only partially. One never knows what kind of emergencies may crop up all of a sudden. Disasters require emergency services at different levels at different times. Even the degrees of resource requirements may vary. As a result, governmental agencies have to be prepared to come up with varying responses, in alignment with the existing situation and circumstances. Therefore, it is impossible to put a lone, one-size-fits-all amateur organization in place for handling every emergency. This can only result in chaos. Therefore, different organizations need to take up different duties. For instance, ARES concerns itself primarily with the activities of non-governmental agencies, like the Red Cross, etc., and with the safety of the local public. It receives its instructions from ARRL. The local leaders of ARES make decisions about the best ways to organize volunteer services, as well as how to interact with all the agencies that they serve. Even local organizations, which are associated with ARES, come forward to help ARES teams train people for emergency services. In case you wish to volunteer, you just have to fill out an online application form. Once you get through, find a local ARES team which will help you to take part in exercises and training. The section manager of the ARRL division in your particular area should help you find an ARES organization. Just look online in order to obtain contact details.

RACES came into existence as a support organization for civil defense. FEMA is its sponsor. Therefore, special FCC rules come into play here, whenever civil defense agencies in the state, local, or county come forward to handle disaster-related services. The state or federal authorities activate such services during civil emergencies. The members of RACES must be members of local groups involved with civil preparedness. Therefore, they undergo special training too. If you want to become a member of this team, you must connect with the local coordinator of Auxiliary Communications Service (ACS). You should be able to find out more about this person from the official RACES website, which offers information about the operations in diverse states. In case you do not find what you are looking for at one website, hunt for others involved with emergency services. Whatever the case, you will do others and yourself a world of good by joining ARES or RACES or both!

MARS

Military Affiliate Radio Services, or MARS, refers to another organization, which comes into play whenever extensive communications are the order of the day during emergencies. In fact, it is the communications link between ham radio operators and global military communications systems. The U.S. Department of Defense is its sponsor in general. However, every branch of the U.S. military operates its own MARS program. The volunteers receive permission to operate, albeit on specific MARS frequencies only, thanks to their special call signs and licenses. These frequencies do not fall into the purview of regular ham bands. This organization offers training in handling technical issues and operations, as well as in being prepared for emergency communications. As a result, volunteers become capable of handling numerous personal messages flowing between people employed in the military and their respective families back home. They may offer to initiate connections between phone systems and radios, or provide phone patches to families. If you wish to join this program, you must be an adult U.S. citizen, holding a valid amateur license.

B. Preparing for Participation in Emergency Operations

Striking up a bond with a single emergency organization, or even a couple of them, is perfectly fine. However, this is just the beginning, for there is more to come. To illustrate, you will need to go through several other steps if you wish to remain prepared to tackle any emergency that may come your way. Towards this end, you will have to become acquainted with the four stages of preparedness: the who, where, what and how.

WHO

You will have to cooperate with the leaders and the particular team belonging to the concerned organization. We cannot mention any particular establishment here, since it is your choice as to which one you desire to link up with as a volunteer. Additionally, you must also interact with the local team leaders and members belonging to your chosen organization. This is because there is no time to think during any kind of emergency. You must have the requisite knowledge and information at your fingertips. In this case, this refers to the call signs that are the creations of the local clubs and stations associated with governmental Emergency Operations Centers or EOCs. Note that time is precious during a time of emergency. Therefore, you must familiarize yourself with their call signs, as well as help the opposite side to become acquainted with your station's call signs. This is possible by participating regularly in exercises and nets. True, you will need to allot some time every week for checking into nets. However, it is imperative that you reinforce your knowledge and awareness about which other ham radio operators in your particular area are participating in your program too. Furthermore, it would be good to set aside some time for attending meetings and other functions. They include work parties, EOC open houses, etc. You will be able to match faces with their respective call signs! Note that having healthy personal relationships in place pays rich dividends whenever a real emergency arises.

WHERE

They have contacted you because it is an emergency. Are you really going to waste time in experimenting with diverse bands, trying to figure out where people are contributing to EmComm? That would truly be ridiculous! Therefore, maintain a detailed list of emergency net frequencies with you at all times. Ensure that the names of the local leaders in various areas are also present on the sheet/chart. If you wish to reduce the size of the list, use a photocopier. Laminate the reduced version, which should now be the size of a credit card. It will last in your purse or wallet for a long time!

WHAT

At the same time, keep your gear prepared for tackling any and every kind of emergency. If an emergency call comes through and you discover that you have to spend some time on getting your equipment together, you will find yourself under terrible pressure. In your eagerness to do everything the right way, you may end up making mistakes. You may even leave some crucial item behind or be unable to find it. The best solution to this kind of confusion would be to have a Go-Kit as an antidote to confusion and chaos.

Go-Kit

A Go-Kit refers to a group of items that you may need in an emergency. You must collect these items in advance and place them in a carrying case. Ensure that the case is not too heavy to handle. This kit will help you to remain prepared for whatever might come your way, without being hassled. You will never need to rack your brain at the last minute, trying to figure out what should actually go into your handy case. At the same time, prior to coming up with your own Go-Kit, give some thought to the kind of mission that may come your way. It could also be that you are planning to participate in more than one emergency mission. Therefore, it would be good to have a detailed checklist. In fact, you might peruse the ARES Field Resources Manual, which gives you some idea of what you should be carrying with you. It is a great generic list, which should prove to be of immense help to you.

In general, your Go-Kit should be able to sustain you for a whole day, an overnight stay or a weekend. Towards this end, ensure that a dual-band or two-meter radio is inside the bag. Along with it, carry some kind of a 'gain' antenna, writing materials, extra batteries and an auxiliary power source. In short, your equipment should suffice to allow you to complete all your communications successfully, wherever you are, without the need for excess transmission power. Furthermore, every item should be lightweight and easy to set up. You must have a copy of your amateur ham radio license and your driver's license on your person too. As for medications, if you feel you will need analgesics and your regular pills along the way, please carry them. Add a basic first-aid kit too.

Water

You may be able to survive without food, but definitely not without water. Therefore, regardless of wherever you go when you are alone, you must have a couple of bottles of water with you. If it is possible, you may carry along a portable filtration system, since you will be able to access running water at various places and be able to drink it safely. In other words, you have access to an unlimited supply of water. Alternatively, you may take along a packet of water purification tablets with you.

Food

It must be things which need no refrigeration at all. It would be good to store non-perishable items for a couple of days or so. Do not believe that you will be able to find something to eat, wherever you go.

Clothing

Check out the weather conditions prior to deciding whether you need warm/cool clothing. It all depends on the kind of emergency that requires your assistance.

References

You will need to keep personal numbers, emergency-related numbers and a list of operating frequencies with you. You never know what you will need and at what time.

Now, it could be that you prefer to operate from home. Of course, then you will not need a Go-Kit. However, you will still have to make some preparations. For instance, you may lose your main antenna, the electricity supply, etc. You might even need a generator in place, for modern radios refuse to be too friendly with batteries. This is because they draw over an amp, during reception. Therefore, a generator proves handy when you may connect it in such a way that the AC circuits within your ham radio shack continue to function properly. In the absence of a generator, you will have to use a back-up power source. You may need to link the DC power supply from your car to your radio. Towards this end, you will have to set up shop in your vehicle. Ponder how you are going to supply power to your equipment, as well as connect an antenna to each part. If you have thought things through before the emergency actually occurs, you should have no issues at all.

HOW

It helps to be completely familiar with all procedures, if you want to be truly prepared for emergencies. Do not bring personal learning or background into the picture at all. You must still have proper training for contributing positively towards emergencies. Towards this end, you may practice with your local EmComm or local NTS. Even the ARRL offers online training courses, for a certain fee. The training includes giving first aid to injured people, cardiopulmonary resuscitation techniques, participating in search and rescue operations, orienteering, etc.

C. Conducting Operations During Emergencies

It may not be possible to adhere to set rules or systematic procedures during an emergency, due to the simple reason that every emergency is different in nature. You will never know what to expect, for things happen all of a sudden! For instance, an earthquake is different from an avalanche, and a wildfire is different from a building going up in flames due to electrical faults. True, you may adopt some common guidelines for tackling emergencies in general, but you will still have to improvise your strategies as you go along.

Follow these principles:

- Whenever there is an emergency, give immediate attention to your family and your property. You owe it to them.
- When you are sure that everyone and everything is safe and secure, volunteer for EmComm services.
- Keep continual track of the primary emergency frequencies on your ham radio.
- The official handling the existing disaster may have some instructions for you. Alternatively, net control might have some tasks for you. You will be able to reach them on that frequency.
- Diligently adhere to orders.
- Whenever they request you to check in, do so. Your superiors will place this request only when necessary.
- It is equally important to keep in touch with the leader who is in charge of local emergency communications. Alternatively, there may be a particular designee handling the same. Always await their instructions.

Now, it is entirely possible that you will encounter some amount of chaos and confusion. This is because everyone is tense, trying to bring the situation under control. Communications may become rather mixed in nature, too. Whatever the case, you must strive to keep a firm head on your shoulders! Just focus on what your training has taught you, such that you prove to be a helping hand, rather than a hindering one. Above all, try to calm others too!

Make a Report

If you think that amateur hams are not involved in reporting accidents, you are badly mistaken! In fact, this kind of reporting is very common, especially when the ham radio operator is driving around. Furthermore, there is no reduction in the frequency of road accidents, fires or stalled cars. Therefore, whenever you come across anything like this, it becomes your responsibility to report it as quickly as possible via your ham radio, especially if it has an auto patch. An auto patch enables you to access an outgoing telephone connection. At the same time, you should have a clear idea of what to say and how to say it to the person listening at the other end. Here are some tips.

- As soon as the first opportunity arises, take your ham radio's power to the utmost limit. Then say 'break' or 'break emergency' in an extremely lucid manner. Even if a lone station is weak, the stronger signal is bound to capture the attention of other listening stations immediately. In fact, do not feel shy or embarrassed if you have to interrupt a conversation in progress. Just go ahead and do your duty.
- As soon as you gain control of the repeater, or you discover that the frequency is clear, declare that you have an emergency to report.
- Give a clear indication that you are about to initiate an emergency auto path. Once you have stated that, activate the repeater's auto patch system without hesitation. In case you find that you are unable to activate it, just request that some other repeater undertake the task on your behalf. If this seems impossible too, find someone on the VHF or HF frequency band. Tell the person at the other end to contact 911 immediately and request that they arrive at the scene of the accident. At the same time, provide all the necessary information to the listener at the other end. After this, do not cut the call. Just remain on standby on the frequency that has connected you to the other person. After all, you want to be sure that your report has gone through. The relaying station must come back and let you know that 911 has heard the distress call and will respond. This will indicate that your call is complete.
- Connect with 911. As soon as you obtain a response from the other end, give your name and state that you are an amateur ham radio operator. Explain that you wish to report an emergency.
- The person at the other end will give some instructions/directions. Follow them. One of them could be a request for you to remain on the line. Do so. Request the other repeater users stay on the line too, as standbys.
- As soon as the operator closes the conversation, you may release the auto patch. Make an announcement about releasing the auto patch.

It is imperative that your communications on your own auto patch, or someone else's auto patch, are always concise and lucid in nature. To illustrate, if you are reporting about a road accident, you must convey information regarding the exact location of the accident, directions to the location, whether the accident is blocking traffic or not, whether people have been injured, if any vehicles have caught fire or have spilled fuel, etc. Do not give elaborate explanations or embellishment of facts. Be brief, for there is no time to spare.

Use an Auto Patch

Repeaters are familiar with this feature. It is possible for a repeater user to initiate a telephone call through another repeater. To illustrate, whenever you access a repeater's auto patch function, you initiate a dial tone. This tone moves through the air. As you continue to move your fingers over the numeric keypad that is an integral part of your microphone/radio, the tones reach a telephone system. In turn, this system dials the number you want and connects you to someone at the other end. Note that every auto patch call is public and on the air. It can never be private in function and is therefore highly useful during emergencies.

You may go in for all kinds of important functions with the auto patch. However, amateur ham radio operators prefer to use it for dialing 911 and reporting accidents only. If you wish to access and activate the auto patch, you will have to use special codes. However, you will need to become a member of the group or club which maintains the repeater if you want to possess these codes. Additionally, the authorities have placed certain restrictions on what you may or may not do with an auto patch. To illustrate, you may not use it for conducting business on any frequency in use by amateur radio operators. You are welcome to order a pizza and perform other personal errands, though! A majority of repeater systems place restrictions on what area codes you may dial, thereby preventing the incurrence of long-distance charges. The best thing about auto patch is that when you are unable to find the correct service provider for your cell phone, or if the pathways on cell-phone systems are overcrowded, it will show you a way through the chaos to reach your destination.

Now, how will you make a phone call with auto patch, after accessing the tone sequences on your radio?

Here are the common steps to follow, although there may be variations between one repeater and another.

- Whenever there is an emergency, do not hesitate to interrupt an ongoing contact/conversation. Then, request that those stations position themselves for an emergency auto patch. Of course, this step will not apply in the absence of any urgency or emergency. Then you will have to wait in line for the channel to clear.
- As soon as you gain access to the repeater, let this person know that you are moving over to auto patch. Towards this end, hold onto the microphone button present on your microphone and do not release it. Next, activate the sequence of tones. Finally, release your hold on the mic's button. To illustrate, your call sign may be NØAX. Declare that NØAX is accessing auto patch. Then press *82 or #11.
- The repeater at the other end will send an acknowledgement. You will know by a unique tone, a telephonic dial tone or a synthesized voice. In case you do not receive an acknowledgement, you will have to repeat the previous step. It could be that your signal is noisy or fluttery. Alternatively, you may just move on to another repeater user and ask this person to activate auto patch on your behalf.
- As soon as the acknowledgement comes through, reach for the tone keys. Dial the number you want, with a dialing code preceding it, if necessary. Of course, this is in accordance with the repeater's special operating requisites.
- Upon receiving an answer to your call, tell the person that you are on radio. To illustrate, give your name and state that you are utilizing an auto patch system on your radio. The person at the other end will realize that both of you cannot talk at once on a radio link. He/she will wait for you to talk first.
- Relay a short and lucid message which is suitable for public hearing.
- When you are through, key in the hang-up tones, if it is a requisite. However, do not take your hand off the microphone's button while doing so. Let the repeater know that you are disconnecting from auto patch.
- You should receive an acknowledgement which is similar to the activation one.
- If you feel it's necessary, inform the other repeater users that they may continue with their normal operations, since you have completed your task.

Make a Distress Call

If someone is in danger and needs immediate help, they can make a distress call on any frequency they have access to. Even before the emergency actually occurs, you should be familiar with the process of handling a distress call on a frequency that is the routine pathway of hammers. This routine pathway could be a repeater frequency covering a large area, a marine service net, etc. Therefore, store one or a couple of such frequencies on your radio's memory. You will not have the time to browse through net directories during an emergency. Then again, it does not matter whether an individual possesses a license or not. He/she may utilize your radio equipment for requesting help during an emergency.

In such situations:

- Give your exact location with other details
- Call out 'mayday, mayday' on a voice mode, followed by "any station, come in, please"
- Describe the issue
- Mention what type of assistance is needed
- Provide any other pertinent information

This means, in order to make a distress call:

- Use the best voice signal of all—MAYDAY, as well as the best Morse code signal of all—SOS.
- After repeating MAYDAY three times, follow it up with your call sign, personal location, location of the emergency, type of disaster, required assistance, etc.
- Go ahead with your distress signal and call sign for a few more minutes. In fact, do not stop until you receive a reply. However, you cannot go on forever. Comfort yourself with the thought that several others may have heard you over your ham radio.
- Change frequencies one after another and repeat when you receive no answer. Make a public announcement whenever you are switching over from one frequency to another. Your listeners will be able to follow you easily.

Sometimes, you are at the receiving end. This means that you can hear a distress call coming over the air. In that case:

- Record all the detailed information coming through in a notebook.
- Take note of the frequency and time of the call, such that you may guide the authorities appropriately.
- Convey everything that you have written in your notebook to the authorities.
- On your part, send a response to the person at the other end. Towards this end, announce your station's call sign, as well as your call sign. Request that the person keep you updated about the existing situation.
- You may even use Morse code to send a message, in order that the sender knows that you can hear him/her.
- Once you have acquired relevant information, request the station in distress to maintain its presence on that particular frequency.
- Call 911 or a public agency. Provide information about the distress call to the dispatcher.
- It may happen that the dispatcher transfers you to a more suitable agency.
- Whatever the case, remain on the air and faithfully adhere to the dispatcher's instructions.
- Keep the station in distress informed about your actions. You are the liaison between the helpers and the station needing help.
- You must remain on the air as long as everyone requires your assistance.

Once you transmit this information, pause for a while to hear an answer. If you do not hear any answer, repeat the information and then pause again. Continue this till you hear a response. The reason the information is repeated each time you transmit is for the listening stations to know the issue and that you are in need of assistance. So never make false distress call or let anyone else do so on your behalf. If this happens, your license can be revoked and you can even be penalized.

On the other hand, if you hear a distress call on any frequency, you can respond. If the distress call is received on a frequency which lies outside the amateur band, for example on the International Marine Distress frequency, ensure you respond only if no other station is responding. For frequencies within the amateur band, you can interrupt the ongoing communication and respond immediately. Listen carefully to what the distress call says and respond accordingly. If you hear the information, let them know that you copy. Try to stay with the station sending the distress call till an action is taken to provide assistance.

Render Support to Outside Communications

It is entirely possible that the disaster/emergency has occurred outside the vicinity of your ham radio station. Is it possible for you to render assistance? Yes, it is! You are welcome to offer your assistance to the communication workers who are on the concerned site. At the same time, step in only when they give you permission or request you to do so. Note that they will keep relaying information, not receiving it. If you enter without warning, you may cause confusion. It is imperative that the concerned authorities receive sufficient knowledge about the disaster, such that they may render appropriate assistance. Of course, there is nothing to prevent you from listening in on the communications.

Whenever an opportunity arises, suggest that you can help. For this to happen, you must tune into repeaters' input and output frequencies. To illustrate, you may discover that there is a station somewhere which is too weak to activate the repeater. You may offer to behave like a relay station via monitoring of simplex frequencies. Declare your call sign and state that you are able to copy both the stations. Do they want you to take over as a relay station? If they agree, go all out to disseminate information out of the disaster site to the requisite authorities.

D. Offering to Serve the Public

We know that hams operate and help during disaster and emergency situation, and the good thing is that such situations are rare. However, as a ham radio operator, you may offer your services for other important public affairs too. This way you may undertake them only when there are no emergencies around to dig into your time.

Hams are therefore occupied with public service communications; they support public events such as sports, races and so on. And they do not wait for an emergency or crisis to happen; they keep themselves engaged in public service communications. This helps not just the public but also the amateurs who want to upskill themselves. However, hams must know that they are not allowed to receive payment in any form for their public services except for the reimbursement for any expenses they make out of pocket, including travel expenses. They cannot charge any sort of fee or ask for something in exchange of this support they provide to individuals or organizations. Many people might wonder why would anyone provide the service for free, however, the reality is that many hams are involved in public service communications on a regular basis, and this might be the reason they even got into ham radio. While providing public service communication, the highest levels of accuracy and performance is expected from the hams, such as:

- They always have to maintain radio discipline—hams should always stick to defined protocols and refrain from irrelevant conversation that hampers public services.
- They cannot become an active part of events—hams can only assist in event communication and cannot participate in events.
- They must maintain their own safety—if they cannot provide services, they are not of any help; that's why they must ensure they are safe and secured.
- They should always protect public information—hams should never share any sort of confidential information while communicating over ham radio.
- They should never go by speculation or guesswork—hams are expected to adhere to the highest levels of accuracy and hence they cannot base their decisions on speculations.
- They cannot give any unauthorized information—if hams are not authorized to see a certain piece of information, they shouldn't seek it out, out of curiousness. They should let the appropriate spokesperson handle it.

A type of communication known as tactical communication is often utilized for coordination purposes, for example—"Go towards the parking area"; "first aid is required on the second floor of the parking deck" and so on. This type of communication is generally handled by using repeater channels or the VHF/UHF simplex systems. Portable, handheld and mobile radios are helpful in specific cases of government services and public safety. To enhance the efficacy and ensure smooth coordination between the stations, the transmitting and receiver side should use tactical call signs that describe a function, organization or specific location, for example—"judges' stand," "school bathroom" and so on. Using these tactical call signs allows the operating stations to change without altering their station call signs; it also frees non-amateur personnel from using amateur call signs. However, these tactical call signs do not stick to the FCC regulations for identifying the stations, although they should be followed. One must still identify himself with his call sign after every 10 minutes in the conversation and before concluding it—as defined by the FCC rules.

Although we've discussed that emergency or disaster communications rarely happen, they are one of the most important reasons we have amateur radios today. This is even mentioned in the purposes listed in the amateur services of the FCC, which places emergency communications as a top priority, over all the other types of communications that are overseen by amateur radio services.

During a serious state of emergency, the FCC can declare a "temporary state of communications emergency," the declaration of which contains special rules or conditions that are to be adhered to at the times of emergency. The state remains active unless the FCC lifts it and the declarations are mostly made through their website, via the ARRL website, through the NTS and Official Relay Stations and via ARRL bulletins. The news is also picked up by the emails that are distributed in the amateur community. The only exception to "nobody owns any specific frequency" on ham bands is during the state of emergency, when you must avoid using the frequency that is being used for emergency or disaster communication.

There is also a need for flexibility that is recognized in case of emergency—in case of threat to property and life. So, if you are in the middle of a disaster, communications are down and you can offer communications to the concerned authorities, you are allowed to do whatever it takes to save life and property. Even medical personnel and public safety personnel are allowed to use your radio for this purpose. So, in case of emergency, if the communication systems are unavailable, you are allowed to use any form to address the risk, and this includes operating outside the frequencies that you are authorized to use. However, you are prohibited from doing one thing—you cannot send information on behalf of your employer or another third party without their consent. This waiver of normal rules is applied only till the emergency state ends. Once conditions are back to normal, one must stick to the normal rules as hams are bound to use and stick to the FCC rules at all times.

Monitor the Weather

Everybody, from A to Z, wants to have prior knowledge about changing weather conditions. Therefore, you may opt for taking up the role of weather watcher. People will welcome your services with a smile, for several areas that encounter terrible weather conditions throughout the year want frequent reports. Of course, the nets, which are responsible for the reporting of weather conditions in local areas, have regular meetings. While some groups prefer to meet at least once or twice each day, others believe in coming together only when adversity threaten to strike. Find out about these groups and learn what you can do. It should be easy to contact them via websites, UHF/VHF repeaters, UHF/VHF 75-meter voice nets, local weather nets, etc.

Attend Sporting Events and Parades

The managers of events feel run off their feet, striving to get everything completed on time! They will definitely welcome your assistance if you promise to keep them updated with the latest information and take up coordination activities. In turn, you will benefit too. Imagine the amount of informal training you will receive via your movements and contacts! You will comprehend communication procedures, as well as learn about operations which are actually simulations of real-life emergencies. To illustrate, visualize a parade as an evacuation process in slow speed. When you take charge of a lost-child booth, are you not engaging in a search-and-rescue operation? Similarly, when you offer to keep track of the entrants participating in marathon races, you are learning about messages dealing with health and welfare. With regard to communications, you will note that the leader of the amateur group takes charge of coordinating plans, along with the managers of the event. After heavy discussions, the group distributes work to all the volunteers. It is possible that all the communicating occurs on a lone simplex frequency, or on several frequencies. Even if these communications concern themselves with mere information related to logistics, status, etc., you will still have learned something new. Thus, the support staff in this communication network works with many needs related to the event.

Now, event managers prefer to work with amateur hams belonging to a single club, or a group of hammers that have come together for a single purpose. Either way, they want them to take charge of the ham radio affairs related to their event.

Therefore, if you would like to make your valuable contribution too, find the local section manager and contact him/her. You are sure to receive this individual's support. Remember that you are going to be present at a parade or a sporting event. Therefore, dress appropriately for the occasion. Ensure that your clothing matches the attire of the rest of the group. If you require special insignia, ensure that you obtain it. Carry your identity card at all times. A copy of your amateur ham radio license and photo proof are requisites too. Sometimes you may be positioned somewhere there is hardly any support. In such a scenario, you had better equip yourself with sufficient food and water. Carry something to protect you against the elements, for weather is always hard to predict. Finally, it would be best to attach your identification to your radio equipment. Alternatively, you may have it engraved on the equipment. This will prevent identity theft.

E. Net Operations

We discussed nets—what they are and what they do. So let's dive deeper and see their different types, structure and other things. Net means a network—these networks help the ham stations to meet on the air and share different kind of information, which is known as traffic. In today's world, where net is associated with the use of computers, a radio net is also comprised of something similar.

Surprisingly, net operations have been in existence for numerous years! In fact, it is quite possible that they came into play even when just two ham radio stations went on air!

Nets are merely scheduled meetings. The participants include those ham radio operators who share similar interests. However, the group does not meet face to face. Instead, it meets on the air. Its interests do not relate only to those things that have a connection with ham radio. They may even include discussions on the Worked All States awards, sharing of information regarding diverse personal collections, invitations to play chess on the air, etc. Some nets believe in being more utilitarian in function. For instance, they prefer to cater to weather reporting services, handling of on-the-air traffic, emergency services, etc. Whatever the agenda, there is no denying that almost all nets function around a similar kind of basic structure.

To illustrate, the Net Control Station, or NCS, launches the net operations, gives direction to various activities, ensures that there is some kind of order and tackles termination of operations in a principled manner. Any individual or group wishing to become part of the net crowd must contact the NCS. NCS stations have their own areas of focus, as well as certain policies in place. The net manager handles this arena, ensuring that net meetings take place on a regular basis, at the same time. Whenever the members feel like exchanging information, they use verbal messaging, or utilize a formal radiogram. The exchange of information takes place during a net meeting, while on frequency or off frequency. The nets, which pay more attention to common interests, or to trading of equipment, prefer transmissions on a single frequency. However, this will not work for groups that focus on handling of traffic, emergency services, etc. Therefore, with the permission of the NCS, such nets go off frequency for communicating with one another and with the public.

Whenever you want to check into a net, ensure that you have registered both your call sign and status/location with the NCS. You must be able to hear the NCS' voice clearly and be able to comprehend the instructions that follow the voice. In case you are not a regular net member yet, wait for the NCS to make an announcement regarding the entry of visitors. As soon as you check in, use the phone, if available, to offer your call sign. In case the NCS is not able to copy your call sign the very first time, you will have to repeat it. Alternatively, the NCS may decide to use a relay from the stations listening into the communication, such that he/she is able to record your call sign. It is possible to check for business (announcements) or traffic (messages) in two ways. One is to let the NCS know that you have a lone item for the net. After acknowledging your item, the NCS will provide further instructions. The second way is to check in using your call sign.

After acknowledging your presence, the NCS will ask if you have any business for the net. Then you may reply that you have one item. In case you want to get in touch with a few other ham radio stations, which are also checking in, make a public announcement about your intention. However, do it in a business-like way, for it must not sound personal at all. After acknowledging your announcement, the NCS will bring you into contact with other stations, following certain procedures. In actuality, it all depends upon how the NCS wishes to operate. The method may adhere to the very formal, or could look like a round-table conference. Whatever the case, it is your responsibility to listen carefully, recognize your NCS and adhere to the directions as strictly as possible. Follow the actions and behavior of other members of the net.

Hams extensively use nets for their communication. These nets are categorized under three types: social, traffic and public service.

Nets are capable of serving small groups and even handle international coverage, formal and informal, etc. Each net is known to have a purpose for its existence and they can be found on both VHF/UHF as well as HF.

Social nets: These are the least formal and the most common type of nets that hams use for their on-the-air meetings. Although these nets can be based on various types of themes, from "stay in contact" nets to hobbies, these are quite easy to join and do not even require any special type of knowledge or training to use. All you need to do is follow what you hear.

Traffic nets: These types of nets are used for exchanging information and routing messages, also known as traffic, to different parts of the world. However, since these are not casual nets, a special structure, known as a National Traffic System (NTS) is defined for them, for effective traffic handling. This structure is comprised of local, national and regional nets during which the traffic is passed from one station to another, until finally, a local net passes it to the defined recipient. These nets function on a regular basis, especially at the time of emergency situations. Although these nets follow a specific procedure, they are easy to join.

Public service nets: These are the nets that come into action whenever they are needed by hams during a state of emergency. Public service nets operate in affected areas where there is a need to monitor weather conditions or other things from time to time to prevent the situation from becoming worse. These types of nets serve a dual purpose—they are used for emergency traffic passing and to coordinate response and reporting activities. However, the training sessions should be attended well in time and not after the disaster strikes. In addition to the emergency services, hams are also involved in communication for public events such as sports, festivals and so on.

After understanding the different types of nets available, it is good to understand their structure. This is because if you want to be a part of these nets, you need to know how these nets are organized, their common practices and signals. If you have a basic understanding of these things, it will help both you and the net.

In the case of emergency nets, the operations are slightly different. Accuracy and efficiency become more important, and the stations need to follow certain principles, known as net discipline, to listen to messages properly and follow the directions. Listening is quite important here because if the important piece of information is not passed as-is, it can lead to issues. Although it is quite natural to add comments and give suggestions during such circumstances, it can lead to mistakes and confusions. Therefore, care must be taken once you enter an emergency net. You must transmit only when you are required to.

Once an emergency net is created, it should always be available to make communications whenever there is a need. Even if there are other important messages waiting in the queue, the preference is always given to the emergency traffic and the precedence is something like this: the emergency messages are passed on first, followed by the priority messages, then health and welfare and later the normal, routine messages.

F. The Handling of Traffic

Before proceeding to learn the tricks of handling traffic, ask yourself if this function is strictly necessary in modern times. After all, there are coast-to-coast links in place, as well as wireless LANs across the globe! True, but handling of traffic still has its own importance. If all the sophisticated communication systems in the world, such as the internet, smart phones, modern computers, etc., were to cease to function suddenly, what then? How would people in various parts of the world communicate with one another? They would have to use the pathways that ham radio amateurs offer!

These amateur traffic handlers will be able to fill in the gap in both accountable and accurate ways, until the sophisticated devices come back into their own once again. This does not mean that modern devices need not be accountable or accurate. However, until they return to their positions, it is imperative to keep the communication channels going loud and strong! For instance, some messages that link to health and welfare are extremely vital and need rapid transmission and receipt. To sum up, apart from keeping the communication pathways flowing well, handling of traffic helps people like you to develop your skills and hone your operating techniques further. In turn, you will never panic when there is an emergency.

Movement of Traffic

The traffic that moves around on the air is comprised of text messages outlined in radiograms. These radiograms are akin to old-style telegrams. In this case, a message moves from the hands of one ham's radio to another, and then a third, and so on, until it reaches its chosen destination. From there, it moves into the hands of the destined receiver. When new entrants to the world of ham radio operations enter the arena, they learn to become responsible, gain experience, take pride in performing well, even under pressure, and develop observation skills. Age, area of residence or work hours do not matter to the authorities. In case you suffer from a physical or visual handicap, do not worry about it either. The authorities just want to prepare volunteers for emergency operations. Now, over time, you may decide that handling of traffic is right up your alley and you wish to continue doing it! Well then, approach your local section manager and request an appointment for Official Relay Station (ORS). If your station receives this appointment, you may go in for specializing in this field via the handling of one mode. Alternatively, you may try to become a net manager, or even take up the role of liaison between two diverse regions.

So you have joined the band of traffic handlers and are eager to go! Your NCS will connect you with another station, after handing you a radiogram for your town. Do not feel bogged down by the quantity of messages flowing through your communication channel. In fact, you may even request that people take it slow, because this is your first day and first message. If you observe experienced traffic handlers, they take their own time, in order to ensure that the job reaches a perfect end.

The Radiogram

The most important task at the time of an emergency is the ability to correctly pass on messages—just the way they are read or heard. Messages are transformed by the radiograms, which are typically comprised of three parts—the preamble, the body and the signature. The preamble is comprised of information regarding the main message. This provides the unique identity for each of these messages so that it can properly tracked until it is received by the recipient. The three parts to a radiogram are:

Preamble

This information consists of—

- A unique number that is assigned to the station originating the radiogram.
- Precedence of the priority—routine, priority, emergency or welfare
- The special handling instructions that are needed to properly handle the radiogram.
- The call sign of the transmitting station that first sends the radiogram.
- A check, which is the number of words and their equivalent in the text of radiogram.
- Place of origin of radiogram
- Time and date of the radiogram
- The complete address where the radiogram is going

Let us say that you have received your first batch of information. This information is telling you something about the message. For instance, it talks about the origin and nature of the message. This is imperative for you to know, such that if there is a need to tackle it and reply, you will be able to do it. This kind of information is the preamble. It is not telling you anything about the message itself. Thus, the preamble takes responsibility for improving the accuracy of any message.

Addressee

This refers to the information that tells you something about the receiver of the radiogram. You will be able to view the name, place of residence and contact details. You will have to deliver the message by phone, or by other suitable methods.

Text Message

Following the preamble is the radiogram's text message. It should be ensured that the receiving station copies the message properly as unusual words can be spelled wrong. Therefore, if need be, standard phonetics should be used for such words. The ARRL has put a limit on the number of words for the text message—it shouldn't exceed a total of 25 words. Although it's not enough to send complete sentences, limiting the words helps in focusing on the key topic and prevention of unwanted words. Also, the longer messages are difficult to relay as they are more prone to errors. Therefore, limiting the message ensures a balance between accuracy and length. Just ensure that the text is free of spelling errors and does not highlight unusual words. An X takes the place of the period, while a question mark shows up at the end of a question. If you look at a check, it refers to a verification of the word count and nothing more. People opt for groups of words, groups of numbers, or groups combining words and numbers.

The text message of the radiogram is followed by the signature, which is usually the name of the person sending the radiogram message. This identification suffices to let you know who sent the radiogram. Whenever you receive a radiogram, spend some time reading everything through from beginning to end. In case there is something that you cannot understand, contact the transmitting station without delay. Only after you are completely satisfied with everything and are sure of delivering the goods, declare QSL. This is the code for 'received completely.' Say it with pride when reporting to your NCS!

If you have been asked to create a radiogram message, you need to follow this format and in case you are not able to decide what to put for the permeable part, you can leave it blank as you might not always have this information, especially in emergency situations. Also, if you do not have the permeable, check with other nets to see if they can help you with the relevant information. If you do not get it from anywhere, the message can be transmitted without permeable information. In case of emergency, the one part that must be included in the radiogram message is the signature—the name of the person originating the message.

The radiogram originated from telegraph messages. The telegraph companies understood the need to track messages and minimize the errors and came up with the concept of adding permeable to all the messages so that it could track the telegraph and reduce the number of errors. And, to their surprise, the concept turned out to be so useful that they decided to take it forward in the world of internet and computer networks. Today, the word permeable is quite commonly used and stands for the same information that it used to carry in the days of telegraph—the information in the heading contains the transmission information about the message for its proper tracking and error minimization. Today, if you look at the Ethernet network, each packet of data is sent along with a preamble that is comprised of a unique number and address, in addition to other things. Even internet messages are secured with a checksum—a word derived from check which is used for error detection. If the message is received with a check different from what was sent in the original message, or if the message is received with a checksum that appears to be different from what the receiver thinks, a retransmission request is raised, considering the fact that the transmission happened with errors.

Deliver the Message

It is very easy. Just call up the addressee and introduce yourself. Do not forget to mention that you are an amateur ham radio operator. State that you have a message from _____. If you do not provide these details, the person at the other end may mistake you for a telemarketer. When your addressee expresses willingness to listen, read out the entire message, albeit clearly and slowly. You do not need to read out the preamble. In case there are numbers in the text, ensure that you know what they stand for, and deliver the message accordingly. You will have to follow this procedure, even if the addressee is another ham radio operator. When you are through, ask the addressee if he/she wishes to forward a reply. Wait until this person outlines the text and provides details of the recipient, his/her address and phone number. Ensure that the words stay within a limit of 25, or even less. If you wish to, opt for numbered radiograms. They will help in saving space. Note down the date and time when you accept the message. Give the radiogram precedence. If you would like to, outline the handling instructions. Finally, count the words. Once you finish recording the message, carry it forward to the next net session. After checking in, state your call sign and declare that you have one message.

As soon as you return to the net, the NCS will allot you a station. This station is ready to accept all your traffic. Therefore, request help for determining the word count of your radiogram. Word count has another name, which is message's check. However, request help only if you genuinely need it. For instance, you may just want a confirmation of the check. It has to be accurate, for if the other forwarding stations discover an error, they will return the radiogram to you. Therefore, whenever you feel that the message is ready for transfer, read it out slowly and clearly to the relaying station in this manner:

- Begin with the preamble and state; break for addressee.
- When the relaying station urges you to go ahead, do so.

- In case he/she asks for clarification related to some item in the preamble, give it.
- Next, read the addressee information aloud. Declare "break for text."
- Read out the text and signature, and state "end of message."

In case the relay station does not comprehend properly, he/she will request that you repeat your readings. The station may use the term "fill" instead of repeat. When the relay station comes back with a QSL, you may experience a sense of achievement!

Chapter 12—Some Handy Tips for Amateurs

A. When you Begin your Journey

It is not easy to develop instant confidence when you are a beginner in this arena. In actuality, you will always wonder what mistakes you are about to make while handling your ham radio station. Therefore, we have come up with certain recommendations for you, which should help you to overcome all those initial glitches and hitches and become a veteran in no time!

You are Not a Know-it-All

Even the most experienced of operators is not perfect! It is not possible for anyone to know everything. Therefore, use all manner of resources, such as catalogs, handbooks, manuals, how-to-articles, websites and magazines. You may also request help from people engaged in 'hammy' discussions on internet forums or clubs. Even a ham on the air will be ready to extend help. Rest assured, no one will refuse to help you. After all, each operator received a helping hand when he/she started his/her business venture.

Be a Good Listener

Talk less, and listen more, for it is the best way to gain knowledge and experience. Ensure that your ears pay close attention to successful ham radio stations. This will help you garner information about successful techniques for operating your own station. It should be easy to find out about real-time seminars related to diverse aspects of ham radio communications. After all, you have access to on-the-air contacts, similar to the manner in which you contact people by email on the internet. Every ham communication is public and open. There is no chance for obscuring or encrypting it.

Be Courteous

Do not be under the impression that you can get away with lack of simple courtesy just because the person at the other end cannot see you or does not know you. Remember, the receiver is also an individual, just like you. Therefore, make terms such as "excuse me," "sorry," "please" and "thank you," integral parts of your behavior and attitude. If the other operator is rude, just switch your attention to someone else. Whatever the case, always keep your mindset flexible and serene while listening and transmitting.

Find a Buddy

Hey, there are several other amateurs like you out there! Why don't you get in touch with someone who is experimenting with a novel venture, just as you are doing? It could be someone who is a member of the same club as you are, or someone from your licensing class. Connect with such people on the air, and even try to use your equipment together. This way you will be able to share in each other's failures and successes, thereby building greater self-confidence along the way.

Join In

So what if you are a newbie? There is no rule stating that operating a ham radio station should be a solitary activity! Therefore, go that extra mile to develop some regular acquaintances. It is fun to receive an on-the-air welcome from members of a local weather net or a roundtable QSO! You will enjoy the fact that each person addresses the other by his/her first name only. No one brings superiority or social statuses into the picture.

Befriend your Equipment

It is not only live people whom you should befriend, but also non-living equipment, if you want to be successful in your business operations. You must get into the habit of turning over the pages of your owner's manual regularly, if you want to know everything about your equipment. In fact, it should become your most important reference book! If your package offers a tutorial or demo, use it! It is imperative that you learn how to handle the settings or main controls efficiently. Similarly, even the most obscure of controls should not defeat you. Therefore, spend time studying your equipment.

Listen to the Manufacturer

Every manufacturing company wants you to benefit from the equipment that it provides you. In fact, if you are satisfied, the company receives the necessary encouragement to improve upon existing tools, gadgets, etc. Therefore, the organization suggests certain types of settings to keep in mind, and procedures to follow, hoping that you will find everything useful. It is imperative that you follow these recommendations, such that you feel comfortable using the gear. Once you are familiar with everything, you should be able to optimize your performance with it.

Risk Experimenting

It is okay to divert from what you are doing, at times. After all, how long will you settle down with the same old band/mode/magazine/radio? As soon as you feel comfortable at your ham radio station, you may move onto something else that captures your fancy. You are bound to find something that will prompt you to take up its in-depth exploration. This way, your creative juices will flow and permit you to do old things in new ways!

Never Let Hiccups Overcome You

It could be that your new antenna refused to work. It could be that the CQ refused to respond to your calls. Never mind! Such hiccups are common. Just do not permit them to overwhelm you or ruin your mood. At the same time, it does no harm to try once more. After all, success is not akin to instant coffee! Note that you have worked diligently for your license. Therefore, do not ever think of giving up!

Be a Relaxed Operator

Are you worried about everyone being able to hear you when you are on the air? Well, you are not the only beginner to face such fears! The majority go through the same trauma. However, what you forget is that it is human to commit errors, especially when you are experimenting with something novel. Treat every mistake as a learning experience. Promise yourself that you will not repeat it, and then discard your guilt and shame from your mind. Determine that you will enjoy being a ham radio operator, regardless of your fumbles and mumbles!

B. Listen to the Experts

You may feel awed by all those veterans who have managed to handle their respective ham radio stations so masterfully! However, they were also novices at one time, just as you are now. Additionally, close research has revealed that they did not opt for high-class ways of operating or any kind of unique formulae. They just relied on simple principles which they could put to work easily, even if they confronted different situations at diverse times. You are welcome to do the same!

Learning is a Continual Journey

You cannot hope to make a perfect start, or even continue with a perfect journey during the initial stages. In fact, you should not be surprised to witness all the errors that you commit along the way. However, you will still benefit, if you make up your mind to treat each problem/issue and goof-up as a learning experience. There is always a next time, wherein you improve upon earlier mistakes. Just do not commit the same mistake repeatedly! Take one day at a time, and you should do fine. Bear in mind that any business venture is an educational journey for a lifetime.

Be a Great Listener

Yes, it is fun to listen to your own voice most of the time! However, try to be quiet and listen more! Paradoxically, a good conversationalist is someone who talks less and listens more. Therefore, every opportunity you get to listen to something interesting from a person you meet, or on the radio itself, pay close attention and absorb knowledge. The masters have gotten where they are purely through the art of listening. Thus, every minute that you utilize for listening becomes a minute of valuable learning and takes you on the pathway to becoming a master yourself.

Details are Important

Every master is well aware that there is a great difference between giving 100 percent to a performance and being almost perfect, that is, giving anywhere between 90 percent and 100 percent to a performance. The same is applicable to being on the air whenever required and not being on the air at such times. In other words, paying attention to the smallest of details always pays off in the end. Therefore, if you feel that your CQ sound is not what it should be, have a professional look at it. Similarly, if you feel that your connector would do better if it had complete waterproofing material in place, go ahead with the job and do not delay. Whatever you do, bear in mind that you are in this field for long-time learning too. Do not neglect anything that might affect your devotion to your task.

Antennas are Important

Do not be stingy when it comes to purchasing good-quality antennas! Just look around you. Have you noticed that top-class ham radio stations spend a big part of their time in getting their antenna systems in place and maintaining their healthy performance? Well, you had better do the same then! At the same time, learn everything you can about building or setting up antennas. Grant importance to feed lines and understand how they work. When you do this, you will discover that you receive more improved returns on your original investment than you would if you were neglecting this vital facet of your radio station. Strive to be a feed line and antenna guru!

Comprehend the Decibels

As in every other business venture, ham radio operators are also in competition with one another. You need to keep one step ahead of the others if you wish to survive in this arena and survive with glory! Towards this end, take note of the signals that you send out via your radio station. They should be easy to understand. However, this will become possible only when the decibels are perfect. They should have strength and be unhampered by disturbances or background noise. In other words, your signals should exhibit great audio quality. The standard of your decibels will decide if the contact wishes to stay with you or leaves you for another radio operator.

Know Your Equipment

Do not continue with your 'I do not know' attitude as far as your equipment and its functioning is concerned! Somewhere along the way, you must learn something about everything! Only when you gain knowledge, however technical and hi-fi it may sound to your ears and brain, will you be able to run an effective and efficient ham radio station. The only concession you can afford is to learn just as much as you need to know, without going whole hog about how electronics and electrical gear operate from A to Z. You must know how to control and how to adjust at all the right times.

Practice, and Practice Some More

Every sharp instrument goes blunt over time. It is the same with sharp brains and minds too! Therefore, it is imperative that you keep practicing with your radio as frequently as you can. In fact, ensure that handling your ham radio station itself becomes an integral part of your day-to-day activities. It should teach you something, as well as help you relax. Always keep yourself updated about changing conditions. If you strive to be as good as your previous masters were, then keep track of which operator is active and when. Know the location of such active operators too.

Regularly Sharpen Your Brain

It is something that experts of ham radio stations have learned from ex-president of the USA, Abraham Lincoln! According to the president, if he had to chop down a tree within eight hours, he would spend only two hours on the actual task. The remaining six hours would go towards sharpening his axe. Well, this is exactly what is required for ham radio operators like you. You have to keep your talents in perfect working order, your brain alert and sharp and your equipment functioning well all the time. This way, should anyone need you on the air, you will be able to handle the assignment wonderfully.

Appreciate Ham Radio History

Ham radio is not a recent invention. In fact, it has a rich and glorious history! You may not need to commit everything to memory. However, it will shape your future behavior if you comprehend how the hobby evolved and shaped itself over the decades. You will be thrilled to acquaint yourself with its origins, as well as the pathways that the evolving methods and conventions have taken. True, they may seem slightly confusing, or even obscure, at times. However, if you peruse the pages of history repeatedly, you should be able to understand everything. In fact, many of these methods and conventions prove useful even in modern times. Strike up an interesting conversation with them!

Do Things the Right Way

There really is no need to feel ashamed of what you do not know. It is not possible for anyone to absorb and remember everything. What should truly worry you is that you may feel you know something, but your knowledge may be all wrong. Maybe you have approached the wrong source or simply haven't comprehended something properly. It could even be that you are only guessing at something, and it is actually wrong. Well, do not be too proud to admit to your mistakes. To err is human. Do not be too proud to approach an experienced person to help you correct your errors and learn to do things the right way. Strive to get your facts right. If you keep ignoring the truth, even if it is in front of your eyes, you will end up making even worse mistakes! Therefore, be bold enough to say "I do not know" when you do not know!

C. Operating Your First Ham Radio Station

You may commit several mistakes during your first attempt at setting up a radio station of your own, filling it up with necessary equipment, budgeting poorly, etc. Your ham radio station may be comprised of just a lone radio, held in your hand, a room at your house, a portable kit or a cell phone. Whatever the case, you may be sidetracked by issues and lose concentration. This may lead to problems in the future too. Therefore, you would do well to consider the tips recommended below.

Getting Used to Used Equipment

It is easy enough to access second-hand stuff. However, you must have a good head on your shoulders, in order to distinguish between genuinely useful and usable equipment and sheer junk! It might help to request that a knowledgeable and experienced friend accompany you, regardless of all your personal expertise. After all, two heads are better than one! Sometimes a deal sounds very good. However, just ask yourself if it sounds so good that it simply cannot be true. Then go ahead with accepting the bargain or rejecting it. Nevertheless, if you manage to obtain good bargains in the used-goods marketplace, you will be saving some money. You may use these savings for purchasing novel bands and modes, should you need them in the future.

Construct Your Own

Building something from scratch is an exhilarating experience! Do you have the requisite skills to cope with small construction projects? If yes, put them to good use at your ham radio station. For instance, your hands may produce keyers, audio switches, filters, etc. True, it does seem easier to purchase new tools and gadgets. However, the expense may burn a great hole in your pocket! Therefore, try to construct some things on your own, with the aid of your soldering iron, drill, etc. There are plenty of magazines, how-to guides, kits and handbooks around the place, to give you the requisite encouragement.

Build Your Own Cables

Any ham radio shack is going to need loads of cables and connectors within and without, thanks to the presence of radio, radio accessories and support equipment. Think of how much you will have to pay if you request professionals take over the connecting and setting up of equipment. Instead, you may obtain pre-made cables on your own, and learn how to connect them to various items. After all, this will not be the first time that you will require cables. As you gain new ideas and implement modifications in your ham radio operating style, you may have to replace old gear, add more equipment, etc. Therefore, give serious thought to budgeting.

Progress Step-by-Step

Do not be in a hurry to obtain everything or do everything. Have patience! It is imperative that you first understand the basic requirements of your shack. You may upgrade your gear step by step, over time. It would be good to have a well-documented plan in place at the initial stage itself. Work out future action strategies, wherein you may fulfill each goal as you gain more experience. It is important that your money and time work in your favor.

Look and Learn

If you can find the email addresses of other ham radio station operators, contact them. They should enjoy answering your questions and providing guidance. After all, few people dislike keen interest and personal attention! Alternatively, if you have their postal addresses, you may write to them too. Check out which stations are located within the nearby vicinity and make a personal visit. If you wish to, you may secure prior permission, although no one will prevent you from entering his or her radio station! Keep a notebook handy, wherein you may jot down relevant advice, handy tips, sudden ideas, etc. Enter thoughts from the websites that you browse, too. Often, the articles are full length, discussing how someone assembled an entire station.

Distribute Your Budget

It is tempting to go for the best-quality radio right away, despite its high price! If you spend everything on the radio, though, what will you do about the rest of the gear that has to accompany it? For instance, you will need cables, antennas and other accessories. In fact, as a novice, you may not even know what exactly you must expend money on, and you will really be done for if you have to keep shelling out money for all manner of extras later on! Therefore, keep a wise head on your shoulders and work out a budgeting plan. Work hard today and you may reap rich rewards later!

Keep Track of the Weakest Link

Your ham radio station will have its own weaknesses, some visible and some hidden. Then again, there is something like the weakest link too, although you may not notice it immediately. Therefore, keep track of where probable points of failure may lie, or what is responsible for loss of quality at your station? You may be encountering certain stations on the airwaves which do not seem to be functioning properly. For example, some exhibit distorted or muffled audio. This is because the station may have a multi-kilo buck radio in place, but a cheap-quality microphone has probably come from a garage sale. Similarly, other stations may have other issues. In your case, if you want to earn a good reputation, remove the weakest link from your station as quickly as possible. Replace it with high-quality gear and ensure that you carefully maintain all the equipment that comes into heavy use.

Be Flexible

Do not launch your ham radio station with the intention of performing the same activities repeatedly for the rest of your life! Towards this end, it would be best not to spend your money on highly specialized gear, which proves good only for performing specific types of operations. You have a computer, as well as accompanying software in place. Use them to perform functions that might change. One example is a digital signal. In other words, make flexibility an integral part of your ham radio operations and personality. Similarly, your rigidity may show up in the way you handle your equipment. For instance, have you nailed everything in place? If yes, you will never be able to move equipment around in future. You may want to have more comfort or make your layout appear more convenient to tackle. There is no denying that the built-in appearance has its own kind of attraction. However, you will not be able to implement any kind of change later on.

Be Comfortable

Do not spend all your time catering to your inanimate objects! Pay some attention to an animated creature too, which is YOU! After ensuring that all your equipment feels comfortable in its allotted place, give some thought to your own comfort. After all, you are going to be spending hours and hours at the operating desk in your shack. Would you not like to settle down in a comfortable chair? Spend some time at stores that sell used office furniture. You may find excellent bargains there. Even the expense will be affordable. Similarly, find desks which are at heights that prove comfortable for your body. Finally, ensure that there is adequate lighting in your shack.

Be Well Grounded

Prior to installing your wires and other equipment in your shack, have a ground system in place. It will be much harder to tackle it later, for you may have to uninstall some things. Do not neglect this aspect. You need good grounding, in order to keep away from RF feedback and ground loops later on. Such issues may prove to be both aggravating and frustrating to tackle!

D. Having Fun on the Radio

Yes, you may have fun in the hours that you devote to sitting at your operating desk. But you are bound to get tired of the day-to-day routine eventually. It may feel like you have gotten into a rut from which there is no escape! Well, there is a solution, wherein you may participate in events and contests.

Listen, and Find Fun

You have access to a radio that connects to anyone and everyone across the globe! Whenever you feel bored, all that you have to is to turn it on. Begin tuning diverse bands and listen carefully. You should be able to find a ragchew which you can join in. Or, you may check into a net and see what is going on. Sometimes, while tackling a slow-scan transmission, you might come across someone discussing or calling a CQ Contest. At other times, a pileup is on display. All that you have to do is dive in! Just think of the vast area that ham radio stations cover and you will comprehend that hundreds of activities must be taking place across the globe. Keep your fingers and ears active and you will be able to participate in something you like, on any given day.

Unique Contests and Events

The weekends are the most active, for this is when people find time for themselves. Remain alert during these days, for you should be able to come across small pockets of activity launching everywhere. You should be able to access the information on VHF or HF bands. Since the initiators would like as many people to participate as possible, they ensure that announcements regarding special events and contests roam the airwaves for quite some time. Furthermore, there is no restriction on who may or may participate. Even a casual passerby in the world of ham radios may become involved. If you make calls to the concerned operators, you will discover how welcoming they are, and how willing they are to give you relevant information. There are also special event stations, which offer interesting and beautiful certificates for contacts.

Create Your Own Contest

Do you have a creative mind which is dying to show off? Well, do so in the form of an innovative contest! It does not matter if the competition is funny, silly or serious. You are just out to have some fun! Invite your fellow club members and friends to join you in thinking up something unique. For instance, get hold of an old bowling trophy. Then create a challenge around it. Set up a contest wherein the person who is able to contact the most states during the weekend wins the trophy. Similarly, set up a radio bingo. You could link up with a friend, deciding to see who between you will be able to make contacts with the least power. This will need tuning up and down the bands, but it should prove to be great fun. These are just a few ideas. You should be able to come up with others on your own.

Join the Parade

There is no denying that people like to enjoy themselves. Towards this end, they organize sports, parades, concerts or festival events. Organizers cannot publicize these affairs without some help. This is where you come in, as an efficient and helpful ham radio operator. They will welcome your help in handling electronic chores and communication details. It should prove beneficial to you too, for many people will encounter your ham radio station and become aware of ham radio operations in the process. They will appreciate the services that ham radios provide. You may contact the local section manager of ARRL and request guidance on how to get in touch with local organizers or even other ham operators offering public services.

Make New Friends

There is no restriction on visiting clubs or nets whenever you want. Apart from gaining new friends, you may develop a completely new outlook on how ham radio operations should take place. Then again, you may decide to acquire knowledge about a subject you never even thought of before. Find a club program that may help you in this regard, and join in. In return for all the kindness and services that you are receiving from your new club, request that members attend your club's meetings too.

Send a Radiogram

There must be people on your list who would love to hear from you on their special days. You have probably been making the excuse that you hardly have time to keep in touch due to your busy schedule. Well, what better time than now is there, when you are getting bored, to take the opportunity to renew that lost contact? Send across a brief radiogram on a relative or friend's birthday. Alternatively, just shout a hello through the radiogram. There should be local NTS nets around the neighborhood which offer radiogram forms. Join in!

Contact Through Satellite

People refer to it as squirting a bird. It is not difficult to make contacts through satellite. To begin with, approach the FM repeater or packet satellites for your task. You may even observe the space shuttle or International Space Station downlink frequencies. In case you are able to hear anything from either one, report it. You will receive an interesting certificate just for your hearing skills! Once you make contact with them, you will love the reflections that move across the heavens at dawn or dusk. In fact, your perceptions about these movements will acquire a completely new meaning and you will never look at them in the same way again!

SWL-ing

This refers to shortwave listening. Consider what exists on the frequencies lying between diverse high-frequency bands. Your radio makes it possible to tune all frequencies, thanks to the presence of general-coverage receivers. Now, when you have such a facility in place, you would do well to experiment with bands and frequencies. For instance, you could go towards the Deutsche Welle, the national broadcaster in Germany. Then again, you could visit HCJB, which is in faraway Quito, in Ecuador, via your radio. Even if they broadcast a majority of programs in the local language, they should have some programs in English, too. You will have to check various websites, though. Alternatively, you may obtain a copy of a shortwave listener's guide. In case you love music, you will be able to listen to anything and everything, without the need for any kind of translation.

Pick Up a New Language

Would you like to learn a foreign language? It could be that you made a beginning when you were in high school, but left it incomplete after leaving school. Well, you have another opportunity to learn something new. Towards this end, contact a DX station. If your equipment works on high frequencies, you should be able to converse with them directly. It is possible to make UHF/VHF contacts with the help of repeaters. You may bring IrLP, or any other kind of linking system, into play. You should find it easy to pick up a new lingo with the help of your ham radio. Begin with a few words each day. Over time, you should be able to grasp an entire vocabulary. This way, you will even gain a couple of DX friends!

Travel to a Cool Place

This is possible if you have portable or mobile ham radio operations in place. You can always travel to an unusual location and use your skills there. For instance, HF frequency affords you an opportunity to function even from a county line, a rural county, etc. In case you have UHF/VHF with a short drive, you can opt for a sought-after grid square. Whatever the case, you are welcome to enjoy a sojourn with Mother Nature, take photographs and even create a novel QSL card for generating contacts.

E. Returning the Favor to the Ham Radio Community

There are so many things that you can do and learn while setting up your ham radio station. Although you enjoy yourself immensely with your ham radio, you can also give back something to the public via a service that uses airwaves. Here are ten ways in which you can make a useful contribution towards the community of ham radio operators.

Be a Tester of Diverse Products or a QSL Manager

Several manufacturers of hardware, as well as authors of varied software, are not completely sure about the efficacy of their products. At the same time, they want to offer only the best to ham radio operators like you. You can help them resolve their doubts by offering yourself as a product tester. After all, who will know better if the hardware or software will benefit ham operations better than an amateur or expert dealing with ham radios? If the answers are in their favor, they can only attract more potential users. Therefore, keep a sharp lookout for requests like these on official websites, blogs and forums. Grab all the opportunities you can to be of service to fellow hammers.

Sometimes, DX stations need assistance in replying to numerous QSL requests. You may offer your services as a QSL manager in this case. You will not only be furthering the sport of DXing, but also gaining international goodwill, through your services.

Engage in On-the-Air Monitoring

Despite all the precautions that amateur bands may take, it is nigh impossible to prevent some, if not many, intruders from entering the airwaves. They are only interested in handling the bands in other people's names, as this allows them to avoid appearing for tests or paying for a valid license. In order to put a halt to such spurious activities, organizations like the ARRL have come up with diverse programs. One of them is Intruder Watch. You may offer your expertise and experience in ferreting out such freeloaders. Become an official observer, wherein you take up the duties of an on-air policeman/policewoman. Similarly, there is another program too. The NOAA has initiated the SKYWARN program, for on-air monitoring. Apart from these well-known establishments, you may find loads of local weather nets looking for hams to volunteer their services.

Be a Representative of Ham Radio

Share your thoughts with representatives in the government, in relation to ham radio operations. There will always be some concerns at the local, state and federal levels. It is imperative that the authorities know about them and take action. Do not give importance only to state and federal problems, for local areas have their concerns too. Your input should prove valuable for arenas like planning, permitting, planning and zoning. Explore all aspects of planned actions too. Some actions could result in adverse effects, instead of beneficial effects. If you alert everyone on time, it will save loads of trouble.

Offer to Experiment

There is nothing to prevent you from experimenting with your creativity. For instance, it could be that you are dying to try out a new design for an antenna which you have figured out all by yourself. Well, go ahead! If the results are favorable, you may share it with others and even ask them for their inputs/opinions. Similarly, you may wish to make further progress in propagation, or even write a simple program. Whatever the case, do not shy away from innovation. In fact, the amateur service encourages people to hop into technical innovation related to radio technology and the techniques to use them. Whenever you do something unique, use magazines and newsletters to let the world know about it.

Be an Elmer

This means that you may join the unofficial group of helpers who lend a helping hand to would-be hams. Towards this end, you may tutor or teach novices all about ham radio operations. Then again, you may aid the person in obtaining his/her first license by doing well in the test. Sometimes, the individual would like to comprehend the rules well. In other words, you take the place of 'friend' in the world of amateur ham radio operations. In fact, there can be no better person to help a newcomer get over the confusing bits of managing a ham radio station.

Be a Volunteer

Every club needs someone who is willing to do something useful and who will work hard. It could be helping with minor services or leadership activities. Whatever the case, the club will definitely appreciate your efforts and time. If you wish to gain new friends, or acquire a new membership, this is the best way to become part of a new family. Never consider any chore too big or small. Handle it because you wish to make a sincere contribution.

Offer Public Service Assistance

It should bring a glow to your heart when you find a rewarding way to contribute to society. Therefore, offer your services as a public service assistant. Once again, approach the local section manager of the ARRL. Request that he/she allow you to take part in a sporting event, fun run, parade, etc. Provide the assurance that things will go smoothly with you in charge. This kind of organization will also provide you with some kind of personal learning. For instance, you will learn how to cope with emergency communications.

Preparedness for Emergencies

Always keep your family and yourself prepared for any kind of emergency that might arise in the future. It is imperative that you grant maximum importance to your family and home during an emergency. When they are safe, you may lend a helping hand to others via your ham radio. After all, if you cannot assist family members first, how can you give assistance to others?

Community Prepares for Emergencies

Join an emergency group in your community, such that you may provide guidance to people in your neighborhood and elsewhere about remaining prepared for emergencies. Remain in touch with the section manager of an organization, or the district emergency coordinator. In case no emergency group exists in your community, you might consider starting one yourself.

Nurture Long-Term Friendships

As a ham radio operator, you have the opportunity to contact people beyond actual physical borders. So reach out to people and bond with them. Strive to build and maintain a dynamic and vital service on your ham radio, such that you find lifelong friends. In fact, you will enjoy your operations more and more when you find other operators and share your experiences. As friends, you may exchange your stories of successes and failures, for they provide valuable lessons. Even if you leave ham radio operations for some time, you are bound to return to the community eventually, because of all the friends that you have found there! You began your venture as a hobby, but now it seems akin to a compulsion!

The Question Pool

Q1: Name the agency that formulates and governs the rules of the amateur radio services in the US?

- a) The FCC
- b) The ARRL
- c) FEMA
- d) ITU

Q2: Which part of the FCC states the rules and regulation around the amateur radio services?

- a) Part 91
- b) Part 97
- c) Part 71
- d) Part 73

Q3: Telecommand signals are the one-way transmissions that are intended to

- a) Terminate the functions of controlled devices
- b) Terminate the functions of uncontrolled devices
- c) Initiate, modify or terminate the functions of controlled devices
- d) Only initiate the functions of controlled devices

Q4: As per the rules and regulations defined by the FCC, which of the following is a purpose of the amateur radio services?

- a) Offering communications for emergency situations
- b) Extending the existing reservoir within the service of trained operators, electronic experts and technicians
- c) Continuation and extension of their unique ability to enhance international goodwill
- d) All the above

Q5: Although no one owns any frequency in the amateur radio service, which type of services are protected against harmful interference as per the rules defined in the Frequency Sharing Requirements of the FCC?

   a) Broadcast services
   b) Radio navigation services
   c) Mobile radio services
   d) None of the above

Q6: Who recommends the transmit-and-receive frequencies for repeater stations?

   a) The FCC
   b) Frequency coordinator
   c) Frequency spectrum manager
   d) None of the above

Q7: What kind of service is allowed on the amateur radio?

   a) Only communicating with other licensed hams around the world
   b) Conducting various kinds of activities and events
   c) Any kind of service that offers an additional source of income to amateurs
   d) Participating in various kinds of radio experiments and communicating with other licensed hams around the world

Q8: Name the agency that works out the international treaties and laws, including the frequency allocations

   a) The United Nations (UN)
   b) The International Telecommunication Union (ITU)
   c) The TCC
   d) All of the above

Q9: Which of the given frequencies is included in the six-meter band?

a) 300 MHz
b) 222.15 MHz
c) 52 MHz
d) 49 MHz

Q10: Which of the given frequencies is included in the 70-centimeter band?

a) 419.50 MHz
b) 222.00 MHz
c) 430.00 MHz
d) 902 MHz

Q11: Which of the given frequencies is included in the 33-centimeter band?

a) 222.15 MHz
b) 300 MHz
c) 904.50 MHz
d) None of the above

Q12: In which of the case base does the amateur radio service fall in the secondary allocation in certain portions of the 70-centimeter band?

a) Digital transmission not allowed in the 70-cm band
b) When they are found to be interfering with important radio-location stations that have primary status
c) US amateurs are given foreign stations in certain portions
d) All of the above

Q13: What should be remembered while operating near the edge of the band assigned to the amateur radio services?

   a) To leave room for the sidebands of the signal
   b) To allow for calibration errors in the frequency display
   c) To leave room for drift in the transmitter frequency
   d) All of the above

Q14: What are the parts of ham bands called where only certain modes can be used?

   a) Sub-bands
   b) Mode-restricted
   c) CW-only sub-band
   d) None of the above

Q15: What sorts of emissions are allowed in the bands at frequencies of 902 and 928 MHz?

   a) Image, phone, data, RTTY, CW
   b) CW only
   c) Digital signals
   d) None of the above

Q16: What sorts of emissions are allowed in the bands at frequencies between 219 and 220 MHz?

   a) CW
   b) Data
   c) Spread spectrum
   d) SSB voice

Q17: Which of the call signs below is a valid US amateur radio?

a) W3ABC
b) KMA3505
c) KMC3506
d) KDKA

Q18: When can you operate in a foreign country as an amateur?

a) Never
b) When the country permits the amateur operation
c) Always
d) Only when you are in mutual agreement with the non-licensed individual of that country

Q19: The FCC requires you to maintain a valid mailing address at all times. If you change your PO Box, what do you need to do?

a) Ensure you update your latest information in the FCC's ULS online system
b) Changes get reflected automatically
c) No need to make any changes
d) Get it updated in their system with the help of a third party

Q20: The FCC normally issues the operator amateur radio license for

a) 6 years
b) Life
c) 10 years
d) 5 years

Q21: How many days after passing the amateur radio license exam can you operate a transmitter on an amateur frequency?

a) After 365 days
b) As soon as the operator license appears in the FCC's license database

c) Immediately after the test
d) As soon as you receive the license via email from the FCC

Q22: Under the vanity call sign rules, who can choose a desired call sign?

a) Any licensed ham
b) Licensed hams with an extra class or general-class license
c) Only those licensed hams with an extra-class license
d) Hams who have had the license for more than 10 years

Q23: Amateurs are allowed to use encryption techniques only in case of

a) Contests and other activities
b) They are always allowed to encrypt the transmitted signals
c) Only while transmitting the commands to radio control craft or space stations
d) Only when operating mobile

Q24: When can amateurs transmit music on amateur radio bands?

a) As part of an authorized rebroadcast of space station transmissions
b) When the music doesn't contain any spurious signals
c) Never
d) When it is meant to interfere with other unauthorized transmission

Q25: Retransmitting the signals from another ham station is generally not allowed. However, it can be done in case of

a) Relaying messages or transmission of digital data from another ham station
b) Can never be done
c) Repeaters
d) Space stations

Q26: What are the three most important elements for an amateur?

a) License, amateur service and operator
b) Operator, station and amateur service
c) FF, operator and amateur station
d) Station, repeater and antenna

Q27: Which important criteria does a candidate need to fulfill other than the test?

a) He shouldn't represent any foreign government
b) He should stay in the country for more than 10 years
c) No other criteria
d) He should know the local language

Q28: When can an amateur station transmit a signal without a control operator?

a) Never
b) In case an automatic control is being used, for example, a repeater
c) When the transmitting station is an auxiliary station
d) When the person who owns the station license is away and another ham is using the station

Q29: Who is responsible for designating the control operator?

a) The station licensee
b) The FCC
c) No one
d) An amateur who has been given the responsibility for ensuring all the FCC rules and regulations are being adhered to

Q30: Who is a control operator?

a) An amateur who ensures all the transmissions comply with the FCC rules
b) A random person who operates all the controls
c) An FFC amateur licensed user
d) None of the above

Q31: Who determines the privileges of an amateur station?

a) The class of operator license that the station licensee holds
b) The frequency designated by the frequency coordinator
c) The topmost class of operator license
d) The class of operator license that the control operator holds

Q32: When is the control operator not needed in case of automatic operation?

a) When the station is transmitting
b) Never required
c) Always required
d) At the time of reception

Q33: which of the systems are allowed to be automatically controlled?

a) Beacons and space stations
b) Repeaters
c) Beacons, repeaters and space stations
d) None of the above

Q34: Who installs the required equipment and procedures for automatic control that is in compliance with the defined FCC rules?

a) Repeater owners
b) FFC itself
c) Control operator
d) Repeater users

Q35: Stations that use a data mode can operate under automatic control in some of the segments, under a frequency above

a) 200 MHz
b) 50 MHz
c) 49 MHz
d) None of the above

Q36: Under which type of control does the APRS network generally operate?

a) Local
b) Automatic
c) Manual
d) Remote

Q37: What type of control is used when the control operator is at the control point?

a) Local
b) Automatic
c) Radio
d) Unattended

Q38: Who owns the responsibility for proper operation of the station?

a) The guest operator
b) The station owner
c) Both of the above
d) Any amateur

Q39: Who owns the responsibility for limiting the access to the station only to responsible licensees who follow the rules and regulations defined by the FCC?

a) The station owner
b) The guest operator
c) Both of the above
d) Any amateur

Q40: What sets the privileges of the control operator? A guest operator hosted by a higher-class licensee can operate using the privileges of the host when

a) The host is the control operator
b) The host is not the control operator
c) The control operator wants
d) Never

Q41: You, holding a technician-class license, are invited to visit the station of your Elmer, who has an extra-class license. You can operate the station on any amateur band and mode when –

a) Your Elmer is away
b) Your Elmer is supervising and acting as control operator
c) Your Elmer runs an errand
d) You can always operate

Q42: You are required to state your call sign

a) To establish a contact
b) Every 10 minutes during a contact
c) Before you conclude the contact
d) All of the above

Q43: In which format can you give the signal?

a) In Morse code only
b) In Morse code, in an image, by voice or as digital transmission
c) In any format
d) In CW only

Q44: Why does the FCC recommend the use of phonetics when identifying by voice?

a) People do not know English
b) It avoids misinterpretation of letters that sound alike
c) Hams like phonetics
d) For hams to learn English

Q45: What are the tactical call signs?

a) These are signs that help identify the location of the calling station
b) These are signs that help identify where a station is and what it is doing
c) They are used to identify the amateur
d) None of the above

Q46: If any station is operating from a location that is in the third district, would he state his call sign as

a) KL333
b) KL7CC/W3
c) KL7W3
d) KL7CC_W3

Q47: How many channels are present in the Citizens Band (CB) for radio communication?

a) 50
b) 40

c) 22
d) 10

Q48: The FCC has set up different types of rules for each type of radio use, and these uses are known as

a) Services
b) Licensing
c) Applications
d) None of the above

Q49: Amateurs give the exams and grade themselves, under the guidance of

a) FCC representatives
b) Other experienced amateurs
c) Volunteer Examiner Coordinator (VEC)
d) National Conference of Volunteer Examiner Coordinator (NCVEC

Q50: Which is the most common method of making contacts, by far?

a) CW only
b) Digital
c) Voice
d) All of the above

Q51: What are the methods called that help develop new ways of converting computer characters into radio signals and vice versa?

a) Protocols
b) Elements
c) Services
d) License

Q52: A radio wave begins its journey with

- a) An antenna as an electrical signal that reverses its direction
- b) A positive electrode
- c) A radio
- d) Special equipment

Q53: What are the hams referring to when they talk about signals?

- a) The electrical energy they use to exchange information
- b) The data signal
- c) The information
- d) None of the above

Q54: What is the signal called that completes a sequence back and forth?

- a) A cycle
- b) Frequency
- c) Wavelength
- d) Time frame

Q55: What do radio designers use to create signals at new frequencies?

- a) Spurious emissions
- b) Unwanted signals
- c) Harmonic signals
- d) Any kind of signal

Q56: What's the range for AM broadcast band?

- a) 88 to 108 MHz
- b) 550 to 108 MHz
- c) 550 to 1700 MHz

d) 900 to 1700 MHz

Q57: At what speed does the radio wave travel through free space?

    a) 88 MHz—108 MHz
    b) At the speed of sound
    c) At the speed of light
    d) The speed increases as the frequency increases

Q58: The formula **wavelength = cycle/frequency** indicates two important relationships between wavelength and frequency, these are

    a) The wavelength decreases with the increase in frequency
    b) Higher frequency takes less time to complete a cycle
    c) Both of the above
    d) Wavelength increases with the increase in frequency

Q59: How can the frequency be converted into approximate wavelength in meters?

    a) The wavelength in meters is equal to 300 divided by the frequency in MHz
    b) The wavelength in meters is equal to 300 multiplied by the frequency in MHz
    c) The wavelength in meters is equal to 300 divided by the frequency in KHz
    d) None of the above

Q60: What is the range of frequency in the VHF spectrum?

    a) 30 to 300 MHz
    b) 30 to 300 kHz
    c) 300 to 3000 MHz
    d) 300 to 3000 kHz

Q61: Why is AM considered to be more inefficient?

a) The carrier doesn't carry any information even when it takes up most of the signal power
b) Each sideband has a copy of the modulating signal
c) Both of the above
d) None of the above

Q62: How does the limiter circuit help in improving the quality of the receiver's output?

a) The limiter strips away all the amplitude variations from the FM signals
b) The limiter adds multiple filters before the output
c) The limiter doesn't help in improving the quality at all
d) The limiter has another device that does the required job

Q63: Which of the emissions below has the narrowest bandwidth?

a) CW only
b) FM voice
c) SSB voice
d) None of the above

Q64: Which of the following are the two components of the radio wave?

a) Voltage and current
b) Ionizing and non-ionizing radiation
c) Alternating current and direct current
d) Magnetic and electric field

Q65: Which of the properties of radio waves listed below is used to determine the various frequency bands?

   a) The magnetic intensity of waves
   b) The time the wave takes to travel one mile
   c) The voltage standing-wave ratio
   d) The wavelength

Q66: Which of the things below might happen when VHF signals are received from a long distance?

   a) Signals get reflected by lightning that takes place in the area
   b) Signals get reflected from the outer space
   c) Signals get reflected from the sporadic e-layer
   d) Signals arrive by ducting

Q67: What is known as sporadic E or E-skip propagation?

   a) When the patches of E-layer in the ionosphere becomes ionized enough to reflect UHF and VHF back to the earth
   b) When irregular propagation happens
   c) When the E-layer is skipped at the time of propagation
   d) None of the above

Q68: What are a meteor and meteor trail?

   a) A meteoroid is also known as a meteor with a trail
   b) The ionized gases in the air are known as meteors and when they become narrow, it is known as a meteor trail
   c) Both of the above
   d) When a meteoroid burns up in the upper atmosphere, it leads to the formation of a meteor; a meteor trail is the ionized gas lasting up to few seconds

Q69: Which of the types of propagation listed below is most commonly seen in over-the-horizon signals on the meter bands of 10, 6 and 2?

a) Sporadic E
b) Backscatter
c) D-layer absorption
d) Gray-line absorption

Q70: Which of scenarios listed below causes troposphere ducting?

a) Temperature inversion in the ionosphere
b) Discharge of lights during the electric storm
c) Tornadoes and hurricane
d) Solar flares

Q71: What do the layers of air with different characteristics form that guides microwave signals for long distances?

a) Tropo
b) Ducts
c) Mobile flutters
d) Tropo contacts

Q72: What causes a high error rate in the digital data signals that are received at UHF and VHF?

a) Multipath
b) Variations in signal strength from multipath
c) Incorrect data transmission
d) All of the above

Q73: What are the attributes that determine which frequencies are reflected best by conductive things in the atmosphere?

a) Variations in the signal strength
b) The wavelength of the radio signal
c) The material of the conductive elements
d) The size and nature of the reflective surface

Q74: What is the range in ionosphere in which the atoms of oxygen and nitrogen gas are exposed to the intense UV rays of the sun?

a) From 30 to 260 miles above the earth
b) From 160 to 260 miles above the earth
c) From 3 to 260 miles above the earth
d) None of the above

Q75: Which are the layers in the ionosphere that refract radio waves?

a) D and E layers
b) E, F1 and F2 layers
c) D, E and F layers
d) D, E, F1 and F2 layers

Q76: What is known as skywave propagation?

a) When radio waves at HF and VHF are completely bent towards Earth, due to refraction in ionosphere layers, as though they were reflected
b) The wave that propagates due to reflection that happens in the layers of atmosphere
c) When a wave travels due to ions present in the sky
d) When radio waves at all the frequencies get reflected from the surface of the earth

Q77: How can the increase in deviation cause interference to the signals on nearby frequencies?

   a) Due to the increased deviation of the nearby signals
   b) Due to increased wavelength
   c) Both of the above
   d) Due to increased bandwidth of the signal

Q78: What is the most common way for amateurs to make long-distance contacts on the HF bands?

   a) Using antennas and repeaters
   b) Sending the signals with high strength
   c) Refraction in the ionosphere layers
   d) A hop that causes the radio waves to be received several miles away

Q79: What is the highest and lowest frequency signal called that can travel between the transmitter and the receiver?

   a) The maximum usable frequency (MUF) and the lowest usable frequency (LUF)
   b) Highest frequency (HF) and lowest frequency (LF)
   c) Maximum usable frequency (MAF) and minimum usable frequency (MUF)
   d) Maximum usable frequency and minimum usable frequency

Q80: What helps in delivering the radio signals to or from the antenna?

   a) The feed point
   b) The wire or cable
   c) The feed line
   d) The resonating antenna

Q81: What are the elements called that are not directly connected to a feed line but that influence the antenna's performance?

a) Driven elements
b) Elements
c) Parasitic elements
d) Driven array

Q82: What is a radio wave made up of?

a) An electric field
b) A magnetic field
c) An electric and a magnetic field
d) There are no fields

Q83: What is polarization?

a) The orientation of the electric field of the radio wave
b) The horizontal orientation of the radio wave
c) The formation of two electric fields
d) The impact of magnetic field of the radio wave

Q84: When an antenna is known to be vertically polarized, how is its electric field aligned?

a) Its electric field is in the direction of the earth
b) Its electric field is aligned horizontally
c) Its electric field is perpendicular to the surface of the earth
d) None of the above

Q85: What is the phenomenon called when the radio waves traveling through the ionosphere undergo a change in their polarization from horizontal or vertical to a combination of the two?

a) Horizontal polarization
b) Vertical polarization
c) Polarization
d) Elliptical polarization

Q86: How does antenna gain impact the signal in a specific direction as compared to a reference antenna?

a) Antenna gain increases the signal strength
b) Antenna gain reduces the signal strength
c) Antenna gain does cause any special impact
d) Antenna gain remains at the same strength as that of the reference antenna

Q87: What is the most common type of pattern that shows the gain of antenna in a horizontal direction around the antenna?

a) An elevation pattern
b) An azimuthal pattern
c) A radiation pattern
d) None of the above

Q88: How much is the characteristic impedance of most coaxial cables that are used in ham radio?

a) 10 ohms
b) 50 ohms
c) 5 ohms
d) 60 ohms

Q89: Which of the options listed below defines the radio horizon?

a) The shortest distance between any two points on the surface of Earth

b) The distance at which any two stations can easily communicate using direct path
c) The farthest point seen when one stands at the base of an antenna
d) The distance between the ground and a horizontally mounted antenna

Q90: What's the band that provides long-distance communications during the peak of a sunspot cycle?

a) 23 centimeters
b) 70 centimeters
c) 10 centimeters
d) All of the above

Q91: Why is ferrite used as a common material in common-mode filters?

a) To reduce RF currents flowing on unshielded wires such as AC power chords, etc.
b) To reduce RF current flowing on the outside of shielded instruments such as microphones, etc.
c) Both of the above
d) None of the above

Q92: What is known as the regulation of the supply?

a) The percentage of voltage change between zero current and maximum current
b) The ratio of maximum to minimum voltage
c) The ratio of maximum to minimum current
d) All of the above

Q93: How can a TNC be connected to the computer's digital data port?

a) USB interface

b) RS-232
   c) COM port
   d) All of the above

Q94: If, instead of microphone input, a sound card is used ...

   a) The output is connected to the TNC
   b) The output of the sound card is connected to the microphone input and the speaker
   c) No kind of interface is needed
   d) The headphone is connected to the TNC

Q95: The RF current in the station can lead to

   a) Erratic operation of computer equipment
   b) Audio distortion, occasional RF burns and erratic operation of the computer equipment
   c) Video and audio errors
   d) Issues in the computer equipment

Q96: What kind of the strap is best for RF grounding?

   a) A solid strap
   b) A twisted pair cable
   c) A round stranded wire
   d) None of the above

Q97: What can be used to fix the distorted audio that is caused due to RF current that flows in the shield of microphone?

   a) Low-pass filter
   b) High-pass filter
   c) Preamplifier

d) Ferrite choke

Q98: The negative-return connection of a mobile transceiver power cable is generally connected to

a) Any metal part of the vehicle
b) The antenna mount
c) The battery
d) The mounting bracelet of the transceiver

Q99: What can happen if the transmitter is working with the microphone gain set too high?

a) The SWR can go down
b) The output power can shoot up
c) The output signal can get distorted
d) The frequency can vary

Q100: What does a squelch control do on a transceiver?

a) It adjusts the automatic gain control
b) It negates the volume of the output noise of the receiver when there is no signal received
c) It sets the transmitter power level
d) None of the above

The Answers

Ans 1: a) The FCC

Ans 2: b) Part 97

Ans 3: c) Initiate, modify or terminate the functions of controlled devices

Telecommand signals are the transmissions intended to initiate, modify or terminate the functions of controlled devices

And 4: d) All of the above

Ans 5: b) Radio navigation services

Since these are crucial, they are an exception as per the rules defined in FCC

Ans 6: b) Frequency Coordinator

Ans 7: d) Participating in various kinds of radio experiments and communicating with other licensed hams around the world

Ans 8: b) The International Telecommunication Union (ITU)

Ans 9: c) 52 MHz

Ans 10: c) 430.00 MHz

Ans 11: c) 904.50 MHz

Ans 12: c) US amateurs are given foreign stations in certain portions

Ans 13: c) To leave room for drift in the transmitter frequency

Ans 14: b) Mode-restricted

Ans 15: a) Image, Phone, Data, RTTY, CW

Ans 16: b) Data

Ans 17: a) W3ABC

Ans 18: b) When the country permits the amateur operation

Ans 19: a) Ensure to update your latest information in the ULS online system of FCC

Ans 20: c) 10 years

Ans 21: b) As soon as the operator license appears in the license database of FCC

Ans 22: a) Any licensed ham

Ans 23: c) Only while transmitting the commands to radio control craft or space stations

Ans 24: a) As part of an authorized rebroadcast of space station transmissions

Ans 25: a) Relaying messages or transmission of digital data from another ham station

Ans 26: a) License, amateur service and operator

Ans 27: a) He shouldn't represent any foreign government

Ans 28: a) Never

Ans 29: a) The station licensee

Ans 30: a) An amateur who ensures all the transmissions comply with the FCC rules

Ans 31: d) The class of operator license that the control operator holds

Ans 32: c) Always required

Ans 33: c) Beacons, repeaters and space stations

Ans 34: a) Repeater owners

Ans 35: b) 50 MHz

Ans 36: b) Automatic

Ans 37: a) Local

Ans 38: c) both the above

Ans 39: a) The station owner

Ans 40: a) The host is the control operator

Ans 41: b) Your Elmer is supervising and acting as control operator

Ans 42: d) All of the above

Ans 43: b) In Morse code, in an image, by voice or as digital transmission

Ans 44: b) It avoids misinterpreting of the letters that sound alike

Ans 45: b) These are the signs that help identify where a station is and what it is doing

Ans 46: b) KL7CC/W3

Ans 47: b) 40

Ans 48: a) Services

Ans 49: c) Volunteer Examiner Coordinator (VEC)

Ans 50: c) Voice

Ans 51: a) Protocols

Ans 52: a) An antenna as an electrical signal that reverses its direction

Ans 53: a) The electrical energy they use to exchange information

Ans 54: a) A cycle

Ans 55: c) Harmonic signals

Ans 56: c) 550 to 1700 MHz

Ans 57: c) At the speed of light

Ans 58: c) Both the above

Ans 59: a) The wavelength in meters is equal to 300 divided by the frequency in MHz

Ans 60: a) 30 to 300 MHz

Ans 61: c) Both the above

Ans 62: a) The limiter strips away all the amplitude variations from the FM signals

Ans 63: a) CW only

Ans 64: d) Magnetic and Electric field

Ans 65: d) The wavelength

Ans 66: c) Signals get reflected from the sporadic e-layer

Ans 67: a) When the patches of E layer in the ionosphere becomes sufficiently ionized, enough to reflect UHF and VHF back to the Earth

Ans 68: d) When a meteoroid burns up in the upper atmosphere it leads to the formation of a meteor and meteor trail is the ionized gas lasting up to few seconds

Ans 69: a) Sporadic E

Ans 70: a) Temperature inversion in the ionosphere

Ans 71: b) ducts

Ans 72: b) Variations in signal strength from multipath

Ans 73: d) The size and nature of the reflective surface

Ans 74: a) From 30 to 260 miles above the Earth

Ans 75: b) E, F1 and F2 layers

Ans 76: a) When radio waves at HF and VHF are completely bent towards Earth due to refraction in ionosphere layers as though they were reflected

Ans 77: d) Due to increased bandwidth of the signal

Ans 78: d) A hop that causes the radio waves to be received several miles away

Ans 79: a) The Maximum Usable Frequency (MUF) and the Lowest Usable Frequency (LUF)

Ans 80: c) The feed line

And 81: c) Parasitic elements

Ans 82: c) An electric and a magnetic field

Ans 83: a) The orientation of the electric field of radio wave

Ans 84: c) Electric field is perpendicular to the surface of the Earth

Ans 85: d) Elliptical polarization

Ans 86: a) Antenna gain increases the signal strength

Ans 87: b) An azimuthal pattern

Ans 88: b) 50 ohms

Ans 89: b) The distance at which any two stations can easily communicate using direct path

Ans 90: c) six or ten meters

Ans 91: c) Both the above

Ans 92: a) The percentage of voltage change between zero current and maximum current

Ans 93: b) RS-232

Ans 94: b) The output of the sound card is connected to the microphone input and the speaker

Ans 95: b) Audio distortion, occasional RF burns and erratic operation of the computer equipment

Ans 96: d) None of the above

Ans 97: d) Ferrite choke

Ans 98: c) The battery

Ans 99: c) The output signal can get distorted

Ans 100: b) It negates the volume of output noise of receiver when there is no signal received.

www.ingramcontent.com/pod-product-compliance
Lightning Source LLC
Chambersburg PA
CBHW081152070526
44583CB00021B/2808